A Concise Introduction to Financial Derivatives

A Concise Introduction to Financial Derivatives seeks to present financial derivatives in a manner that requires minimal mathematical background. Readers will obtain, in a quick and engaging way, a working knowledge of the field and a collection of practical working insights. The book is ideal for aspiring young practitioners, advanced undergraduates, and masters-level students who require a concise and practice-led introduction to financial derivatives.

Features:

- Practical insights and modelling skills.
- Accessible to practitioners and students without a significant mathematical background.

Eben Maré holds responsibility for absolute return portfolio management and has been working in the financial markets for the last 33 years. He has also held senior roles in risk management, treasury, derivatives trading, and asset management. He has a PhD in Applied Mathematics and is an associate professor in Mathematics and Applied Mathematics at the University of Pretoria in South Africa. He has wide research interests in financial derivatives, asset management, and financial markets.

A Concise Introduction to Financial Derivatives

Eben Maré

CRC Press is an imprint of the
Taylor & Francis Group, an **informa** business
A CHAPMAN & HALL BOOK

Designed cover image: © ShutterStock Images

First edition published 2025
by CRC Press
2385 NW Executive Center Drive, Suite 320, Boca Raton FL 33431

and by CRC Press
4 Park Square, Milton Park, Abingdon, Oxon, OX14 4RN

CRC Press is an imprint of Taylor & Francis Group, LLC

© 2025 Eben Maré

Reasonable efforts have been made to publish reliable data and information, but the author and publisher cannot assume responsibility for the validity of all materials or the consequences of their use. The authors and publishers have attempted to trace the copyright holders of all material reproduced in this publication and apologize to copyright holders if permission to publish in this form has not been obtained. If any copyright material has not been acknowledged please write and let us know so we may rectify in any future reprint.

Except as permitted under U.S. Copyright Law, no part of this book may be reprinted, reproduced, transmitted, or utilized in any form by any electronic, mechanical, or other means, now known or hereafter invented, including photocopying, microfilming, and recording, or in any information storage or retrieval system, without written permission from the publishers.

For permission to photocopy or use material electronically from this work, access www.copyright.com or contact the Copyright Clearance Center, Inc. (CCC), 222 Rosewood Drive, Danvers, MA 01923, 978-750-8400. For works that are not available on CCC please contact mpkbookspermissions@tandf.co.uk

Trademark notice: Product or corporate names may be trademarks or registered trademarks and are used only for identification and explanation without intent to infringe.

Library of Congress Cataloging-in-Publication Data

Names: Maré, Eben, author.
Title: A concise introduction to financial derivatives / Eben Maré.
Description: First edition. | Boca Raton : Chapman & Hall, CRC Press, 2025.
| Includes bibliographical references and index.
Identifiers: LCCN 2024022717 (print) | LCCN 2024022718 (ebook) | ISBN 9781032630854 (hbk) | ISBN 9781032637082 (pbk) | ISBN 9781032637099 (ebk)
Subjects: LCSH: Derivative securities.
Classification: LCC HG6024.A3 M3465 2025 (print) | LCC HG6024.A3 (ebook) | DDC 332.64/57–dc23/eng/20240705
LC record available at https://lccn.loc.gov/2024022717
LC ebook record available at https://lccn.loc.gov/2024022718

ISBN: 978-1-032-63085-4 (hbk)
ISBN: 978-1-032-63708-2 (pbk)
ISBN: 978-1-032-63709-9 (ebk)

DOI: 10.1201/9781032637099

Typeset in Minion
by SPi Technologies India Pvt Ltd (Straive)

To my beloved Amanda. Thank you for being there for me.

Contents

Preface, xiii

Acknowledgements, xvi

List of Symbols, xviii

List of Abbreviations, xix

Chapter 1 ▪ Markets	1
1.1 INTRODUCTION	1
1.2 EXCHANGES	4
1.3 EQUITY MARKETS	7
1.4 COMMODITY MARKETS	8
1.5 CURRENCY MARKETS	11
1.6 RATES MARKETS	13
1.7 DERIVATIVES MARKETS	14

Chapter 2 ▪ Market Players	18
2.1 INTRODUCTION	19
2.2 SPECULATORS	19
2.3 INVESTORS	20
2.4 HEDGERS	22
2.5 ARBITRAGEURS	23
2.6 MARKET MAKERS	26
2.7 HEDGE FUNDS	27
2.8 CENTRAL BANKS	29
2.9 REGULATORS	34

viii ■ Contents

Chapter 3 ■ Rates	37
3.1 INTRODUCTION	37
3.2 RATES	38
3.3 BONDS	38
3.4 PRACTICAL ISSUES	40
3.5 FURTHER READING	40

Chapter 4 ■ Derivatives	41
4.1 INTRODUCTION	42
4.2 FUTURES AND FORWARDS	42
4.3 OPTIONS	43
4.4 REAL OPTIONS	45

Chapter 5 ■ Option Strategies	48
5.1 INTRODUCTION	49
5.2 VANILLA OPTIONS	50
5.3 ABOUT VOLATILITY	51
5.4 TERM STRUCTURE OF VOLATILITY	53
5.5 STRATEGIES	55
5.6 PACKAGES	57
5.7 RISKS AND CONSIDERATIONS	57
5.8 FURTHER READING	59

Chapter 6 ■ Basic Option Bounds	60
6.1 INTRODUCTION	60
6.2 BASIC ASSUMPTIONS	61
6.3 ELEMENTARY RELATIONS: CALL OPTIONS	62
6.4 EUROPEAN PUT–CALL PARITY	66
6.5 ELEMENTARY RELATIONS: PUT OPTIONS	67

Chapter 7 ■ Relations Between Options	70
7.1 INTRODUCTION	71
7.2 STRIKE PRICE RELATIONS: PUT AND CALL OPTIONS	71

7.3	AN APPLICATION OF BUTTERFLY SPREADS	72
7.4	REPLICATING GENERAL EUROPEAN PAYOFFS	74

CHAPTER 8 ▪ Binomial Pricing Model: I — 76

8.1	INTRODUCTION	77
8.2	HOW DO WE PREVENT ARBITRAGE?	77
8.3	BINOMIAL OPTION PRICING	79
8.4	DISCUSSION OF ASSUMPTIONS	81
8.5	CHOICE OF PARAMETERS	82
8.6	RELATIVE ASSET PRICES	83
8.7	LIMITING BEHAVIOUR	85
8.8	FURTHER READING	86

CHAPTER 9 ▪ Binomial Pricing Model: II — 87

9.1	INTRODUCTION	88
9.2	BINOMIAL ARBITRAGE EXAMPLE	88
9.3	AN EXAMPLE OF PUT–CALL PARITY VIOLATION	90
9.4	ALTERNATIVE PARAMETER DERIVATION	91
9.5	BINOMIAL METHOD EXAMPLES	91
	9.5.1 Methodology	91
	9.5.2 European Put Option	92
	9.5.3 American Put Option	93
9.6	BINOMIAL CONVERGENCE	95
9.7	FURTHER READING	95

CHAPTER 10 ▪ Option Values: I — 98

10.1	INTRODUCTION	99
10.2	LOG-NORMAL PRICE DYNAMICS	99
10.3	EXPECTATIONS	100
10.4	EUROPEAN CALL VALUE	101
10.5	FORWARD STARTING OPTION	103
10.6	OPTIONS ON FORWARDS/FUTURES	104

10.7 APPROXIMATE CALL VALUES	105
10.8 MONTE CARLO SIMULATION	106

Chapter 11 ▪ Option Values: II — 110

11.1 INTRODUCTION	111
11.2 CONTINGENT CLAIM ANALYSIS	111
11.3 DISCUSSING THE HOLES IN BLACK–SCHOLES	112
11.4 ADJUSTMENT FOR CONTINUOUS DIVIDENDS	116
11.5 INVOKING TRANSACTION COSTS	117
11.6 CONSTANT PROPORTION PORTFOLIO INSURANCE	118
11.7 INCLUSION OF COLLATERAL IN THE MODEL	119
11.8 FURTHER READING	121

Chapter 12 ▪ Black–Scholes PDE — 122

12.1 INTRODUCTION	123
12.2 GOVERNING PDE	123
12.3 INITIAL BOUNDARY-VALUE PROBLEM	124
12.4 SOME INTERESTING SOLUTIONS	124
12.5 DUAL BLACK–SCHOLES PDE	128
12.6 EXAMINING A SPECIAL CASE: $\sigma = 0$	129
12.7 REDUCTION TO THE HEAT EQUATION	130
12.8 FURTHER READING	131

Chapter 13 ▪ Perpetual Options — 132

13.1 INTRODUCTION	132
13.2 GOVERNING DIFFERENTIAL EQUATION	132
13.3 PERPETUAL EUROPEAN CALL	133
13.4 PERPETUAL EUROPEAN CALL WITH A DOWN-AND-OUT BARRIER	134
13.5 PERPETUAL AMERICAN PUT	135
13.6 COMMENTS AND FURTHER READING	136

Contents ■ xi

Chapter 14 ■ Application: Corporate Credit	137
14.1 INTRODUCTION	137
14.2 CAPITAL STRUCTURE OF THE FIRM	138
14.3 RECOVERY	140
14.4 PRACTICAL APPLICATION AND LIMITATIONS	140

Chapter 15 ■ Greeks	143
15.1 INTRODUCTION	144
15.2 BASIC FORMULAE	147
15.3 DERIVING SENSITIVITIES	147
15.4 BASIC GREEKS	149
15.5 PRACTICAL ASPECTS	150

Chapter 16 ■ Exotic Derivatives	152
16.1 INTRODUCTION	152
16.2 WHAT DO I GET, AND WHEN?	153
16.3 EXOTIC VARIETIES: THINKING ABOUT PRODUCTS	155
16.4 SPOT-RELATED CHANGES	156
16.5 STRIKE-RELATED CHANGES	156
16.6 CHANGES AFFECTING THE OPTION MATURITY	158
16.7 MULTI-ASSET OPTIONS	161
16.8 REMARKS AND FURTHER READING	163

Chapter 17 ■ Model Validation Process	166
17.1 MOTIVATION	167
17.2 INTRODUCTORY EXAMPLE	167
17.3 SCOPE	170
17.4 MODEL VALIDATION	172
17.5 VALIDATION METHODOLOGY AND PROCESS	175
17.6 CONCLUSION AND FURTHER READING	178

Chapter 18 • Risk	180
18.1 INTRODUCTION	181
18.2 RISK MANAGEMENT	184
18.3 RISK MANAGEMENT PROCESS	186
18.4 POTENTIAL PROBLEMS	189
18.5 CONCLUSION	189

REFERENCES, 190

INDEX, 195

Preface

> *We have a habit in writing articles published in scientific journals to make the work as finished as possible, to cover up all the tracks, to not worry about the blind alleys or describe how you had the wrong idea first, and so on. So there isn't any place to publish, in a dignified manner, what you actually did in order to get to do the work.*
>
> – RICHARD FEYNMAN (1918–1988),
> 1966 NOBEL LECTURE

Numerous books have been written on the subject matter of financial mathematics and financial derivatives. These books typically range very widely in their level of technical difficulty and comprehensiveness. Why another?

Richard Hamming famously wrote:

> The purpose of computing is insight, not numbers.

Correspondingly, our aim here is to achieve accessible insight with the least amount of mathematical obfuscation.

We have entered an environment where information has become 'cheap' and easy to access. The value we add, in practice, will therefore be in our ability to use the information, to reason about it, and to focus on the various insights these provide. The quote by Feynman is relevant in this regard – we need to be willing to play around and attempt various solutions to problems. It is also important to attack problems from different angles to ensure understanding, insight, and correctness.

With this in mind, our thinking has been to keep the mathematical sophistication to a minimum while trying to showcase, and introduce, some of the thinking behind so-called derivative markets.

The book is intended for a wide audience with a basic level of mathematical familiarity. Individuals with a knowledge and understanding of calculus will find the material accessible.

The chapters of the book proceed as follows. Chapters 1 and 2 provide an overview of the markets and the different players in the markets. It is important for anyone interested in this field to be acquainted with the players, who determine the price action in the market. An understanding of the different financial markets and how they fit into the bigger picture is highly desirous. It is important to understand some of the 'jargon.'

Chapter 3 provides a short introduction to interest rates. The cost of money is a basic ingredient of derivative replication and pricing. Chapter 4 introduces the concept of derivatives while the focus of Chapter 5 is on options with some practical risk considerations. We also have a discussion on the concept of volatility and its applications.

The basic modelling starts in Chapter 6 with an overview of option bounds, which are derived using the *law of one price*, specifically applied to European options. Chapter 7 is devoted to butterfly strategies and demonstrates the concept of static option replication as well as a discussion on risk-neutral pricing.

The binomial option pricing method is introduced in Chapter 8 and covers the basic derivation with an application to show that the method is consistent with the Black–Scholes PDE in the limit as the time-step tends to zero. The binomial method is a great pedagogical tool and provides a vital example where pricing and hedging, specifically in a derivatives context, go hand-in-hand. In Chapter 9, we highlight some applications of the binomial method.

In Chapter 10, we provide an intuitive explanation of log-normal asset price dynamics and use that concept to derive the price of a European option. We also use the opportunity to introduce the Monte Carlo simulation method. Chapter 11 demonstrates consistency with the Black–Scholes framework, with the focus on model-based assumptions, and provides some extensions of the basic framework.

In Chapter 12, we cover the Black–Scholes PDE (with a third derivation) and some special solutions. Chapter 13 is devoted to the special case of perpetual options.

In Chapter 14, we show an application of derivative pricing technology to the pricing of corporate credit. Hedging derivative securities is of vital concern. In this context, we introduce the option 'Greeks' in Chapter 15 with a practical discussion of hedging. In Chapter 16, we extend the basic ideas behind derivative instruments to so-called exotics.

Chapter 17 discusses the important concept of model validation with application to derivative securities. The principles, generally, apply to any mathematical model. Chapter 18 covers a short discussion on risk management, with the focus on market risks.

Disclaimer: In this book, I have tried to provide a brief, concise overview of derivatives. The field has grown to such an extent that this task was almost impossible. I might not have given due credit to the pioneers in various sub-fields, and there are also numerous, relevant, and additional references that I could have added; my apologies, it is not my intention to offend anyone. Our strategy, throughout, has been to provide insight. This book is meant to serve as an introduction to a field I love, and I have taken great care in proofreading it. However, I disclaim any responsibility for any remaining errors; further, I accept no liability arising out of, or in connection with, the use of this book.

Acknowledgements

I am grateful to:

- My family for their support, especially my wife Amanda, my children Eben, Jean-Jacques, Desiré, Yolandi, Carien, and my grandchildren Jordan, Chris, Mark, Aurora, and Desirée, as well as feathered child, Jessica, who shared in the delights (and miseries) of this project.

- Previous and current colleagues and friends at Nedbank, FirstDerivatives, Gensec, Nedcor Investment Bank, STANLIB asset management, and Absa asset management. In particular, Graham Smale, Francois Oosthuizen, Anton Botha, Chris Doyle, Peter Lane, Warren Burns, Carolus Reinecke, DeClercq Wentzel, Moepa Malataliana, George Brits, Tebogo Tsotetsi, Patrick Mmamathuba, Bhekinkhosi Khuzwayo, Alida Herbst, Gareth Witten, Armien Tyer, Kurt Benn, and Kanyisa Ntontela.

- Colleagues at the University of Pretoria, for support. Especially, Mapundi Banda, Gusti van Zyl, Dries van Niekerk, Rodwell Kufakunesu, Brenda Wingfield, Jean Lubuma, Conrad Beyers, Anton Ströh, Gary van Vuuren, and the late Yvonne McDermott.

- I received suggestions and comments from many colleagues and friends. Kind thanks to Leon Sanderson, and also, alphabetically, kind appreciation to Christina de Klerk, Emlyn Flint, Alexis Levendis, Byran Taljaard, Pierre Venter (who also assisted me with some figures), and Vaughan van Appel. Desiré Maré assisted greatly with proofreading.

- Kind appreciation to colleagues who have guided me over time – Jonathan Ziveyi, John O'Hara, Edson Pindza, Dave Bradfield, Brett Dugmore, Yashin Gopi, Riaan de Jongh, and Mike Sherring. My sincere debt and gratitude to my late friend and mentor, Schalk Schoombie.

- The kind staff from Taylor & Francis, especially Callum Fraser and Mansi Kabra for their sincere support and patience. I appreciate the assistance from Divya Muthu from Straive as well.

- In the mid-nineties, I had the privilege of meeting Peter Carr and Emanuel Derman at their respective offices. Despite their busy commitments, their energy, enthusiasm, and passion for derivatives was contagious. I wish to thank them for their kind guidance.

SOLI DEO GLORIA

List of Symbols

$a \approx b$	a is approximately equal to b
$a := b$	a is defined as b
$a \equiv b$	a is equivalent to b
$a \ll b$	a is much less than b
δt	period of time
$\frac{d^k f(x)}{dx^k}$	k-th order derivative of $f(x)$
$f'(x)$	derivative of $f(x)$ w.r.t. x
∂_x^p	partial derivative of order p with respect to x
∂_x^p	$\frac{\partial^p}{\partial x}$
$\partial_x f$	$\frac{\partial f}{\partial x}$
$S(t), S_t$	spot price at time t
$C(S_t, X, T)$	call option at time t with underlying S_t, strike X, and maturity T
$NN(\cdot)$	the standard cumulative normal distribution function
N	number of degrees of freedom in space discretisations
X^+	positive value of X : $X^+ = \max(X, 0)$
$B(t, T)$	value of zero coupon bond at time t, which pays 1 at time T
Π_j	product over j terms
Σ_j	summation over j terms
$N(\mu, \sigma^2)$	the cumulative normal distribution function with mean μ and variance σ^2.

Remark: We introduce relevant notation in different sections, this is not meant to confuse as we frequently suppress variables which are not directly relevant.

List of Abbreviations

BOE	Bank of England
CBOE	Chicago Board of Options Exchange
CBOT	Chicago Board of Trade
CDS	Credit default swap
CME	Chicago Mercantile Exchange
DJIA	Dow Jones Industrial Average
ECB	European Central Bank
ETF	Exchange traded funds
FD	Finite difference(s)
FED	Board of Governors of the Federal Reserve System
IVP	Initial value problem
ICE	Inter Continental Exchange
LME	London Metals Exchange
LSE	London Stock Exchange
NYSE	New York Stock Exchange
NYMEX	New York Metals Exchange
ODE	Ordinary differential equation
OTC	Over-the-Counter
PDE	Partial differential equation
P/L	Profit/Loss
PVBP	Present value of a basis point
UST	U.S. Treasuries
ZCB	Zero-coupon bond

CHAPTER 1

Markets

Remember the Golden Rule: He who has the gold, makes the rules.

– PETER CARR (1958–2022)

Markets can remain irrational longer than you can remain solvent.

– JOHN MAYNARD KEYNES (1883–1946)

The market is a complex adaptive system, and it's best viewed as a collection of diverse opinions, not an absolute truth.

– ADAM NASH

Price is a creature of the market's mood. In booms, it is set by the greediest buyer; in busts by the most fearful seller.

– BENJAMIN GRAHAM (1894–1976)

1.1 INTRODUCTION

We have all seen real-time price quotes, from major financial service providers, scrolling across television screens, which detail equity, currency, commodity, and, even, crypto-currency prices. We have witnessed financial market prices react to news events such as earthquakes, floods, or geopolitical matters, for example.

Andrew Lo, the author of *Adaptive Markets* [69], writes:

> Stock markets are merciless in how they react to news. Investors buy or sell shares depending on whether news is good or bad, and the market will incorporate the news into the prices of publicly traded corporations.

This statement is corroborated by legendary economist John Maynard Keynes, who famously reflected on the dynamics of markets, as follows:

> The markets are moved by animal spirits, and not by reason.

Financial markets reflect perceptions of the aggregate view of the various participants on 'issues of the day;' anything with bearing on the value of a security can lead to dramatic price fluctuations. Revered investor Benjamin Graham remarked:

> In the short run, the market is a voting machine. In the long run, it is a weighing machine.

Indeed, esteemed economist Joseph A. Schumpeter said:

> There exist no more democratic institution than the market.

Therefore, financial markets matter! But our desire for trading has been around 'forever.' This humorous anecdote from Louis Gave [49] expands on that notion:

> In Alaska, during the first gold rush, one winter was particularly rough and a famine ensued. To survive, people only had sardine cans, and a lively market took place in this rare commodity. One fellow bought himself a can of sardines at an extraordinary price, but was surprised to find, upon opening the box, that the sardines were rotten. He went back to complain, but was told by the can's previous owner: "but those weren't eating sardines, they were trading sardines!"

On a broad conceptual level, we can view the financial markets as a dynamic and multifaceted entity that plays a central role in the global economic landscape. It serves as the arena where buyers and sellers engage and converge on the amount and price for goods, services, products, and financial instruments. As Mauboussin [78] points out:

> So when you're dealing in markets, it's not enough to have your own view, you have to consider what other people think.

At their core, all markets operate on the basic economic principles of supply and demand. This concept dictates that prices and quantities of goods and services are determined by the continuous interaction and interplay between buyers and sellers. Fundamentally, when demand exceeds supply (i.e., buyers dominate), prices tend to rise, which incentivises producers to increase output. Conversely, when supply outstrips demand (i.e., sellers dominate), prices decrease, which prompts producers to adjust their production levels. This delicate balance is the heartbeat of the market, which drives economic and financial market activity while shaping the allocation of resources.

Eugene Fama notes the delicate risk versus return balance, which forms the basis of our discussions:

> Markets are efficient, but there are different dimensions of risk and those lead to different dimensions of expected returns. That's what people should be concerned with in their investment decisions and not with whether they can pick stocks, pick winners and losers among the various managers delivering basically the same product.

The financial markets are our laboratory, we wish to create products using the markets and find ways to price and hedge them. We, therefore, need to start by understanding the financial markets. Jan Loeys [70] provides some well-earned insight:

> We should start from the assumption that the market reflects the collective weighted opinion of millions of investors across the world who have much at stake in avoiding being wrong. This is better than to assume that there is a lot of irrationality, emotions, and behavioral biases in markets. I am sure these are there, but they are not the right starting point to consider what to do in markets. I always start from the assumption that the

market is right as it reflects the wisdom of the crowds. Only then should one start to probe prices and see whether there are some obvious holes in them. Overconfidence that we know better than the collective wisdom of everyone else has been the downfall of many an asset manager and strategist, and I have felt its brunt many times.

In the next sections, we will consider different financial markets. In this work, we frequently focus on examples in equity-related markets; however, the principles apply to currencies, bonds, and commodities as well (sometimes with some modifications). For the purpose of our introduction, however, we shall briefly discuss all the relevant financial markets.

We start by looking at financial exchanges, which serve as the delivery mechanism of formalised financial market transactions and instruments.

1.2 EXCHANGES

Economies are based on trade. People trade and exchange goods for services.

Central to understanding modern financial markets, and their various participants (see, Chapter 2), is an understanding of financial exchanges. Financial exchanges provide a centralised means of trading standardised products. This serves to facilitate liquidity and has been a significant contributor to the mainstream success of modern financial markets.

Financial exchanges date back centuries and have evolved over time; they have been shaped by changing economic conditions, market needs, and regulatory developments. The ever-changing economic conditions have shaped their history into a story of evolution, innovation, and adaptation.

Financial exchanges have played a crucial role in facilitating the buying and selling of financial instruments, which, in turn, provides liquidity and contributes to the overall functioning of the global financial system. A brief history of equity market exchanges is as follows:

1. **Amsterdam Stock Exchange (1602):** Often considered one of the earliest examples, the Amsterdam Stock Exchange, established in 1602, facilitated the trading of shares of the Dutch East India Company. This marks the formalisation of a marketplace for buying and selling financial securities.

2. **London Stock Exchange (1698):** The London Stock Exchange (LSE) was founded in 1698 under the Buttonwood Agreement, making it one of the oldest stock exchanges in the world. It initially operated from coffee houses, but later moved to a more formal setting on Capel Court.

3. **Paris Bourse (1724):** The Paris Bourse was established in 1724, and it played a pivotal role in the development of financial markets in France. Over time, it became an essential part of the European financial landscape.

4. **New York Stock Exchange (NYSE) (1792):** The New York Stock Exchange was founded in 1792 when 24 stockbrokers signed the Buttonwood Agreement under a buttonwood tree on Wall Street. Schwed [99] describes it as follows:

   > The genesis of Wall Street was a buttonwood tree under which buyers and sellers used to meet. That tree perfectly fulfilled the pure function of a market place; it was a known spot where a man could go to do financial business. A necessary code of procedure for trading was recognized.

   The NYSE has since become one of the world's largest and most influential stock exchanges.

5. **Tokyo Stock Exchange (TSE) (1878):** The Tokyo Stock Exchange, founded in 1878, played a significant role in the economic development of Japan. It grew in importance as Japan emerged as a major global economic player.

Our main focus in this book is on derivatives and their applications, as well as aspects pertaining to their pricing and hedging.

One can consider futures contracts as one of the most basic derivatives. (See, Chapter 4.) The origins of futures trading can be traced back to the 17th century in Japan. Rice futures were traded on the Dojima Rice Exchange in Osaka (see, [25]). These early futures contracts were created to help farmers and merchants manage the risks associated with fluctuations in rice prices. Futures trading gained momentum in Europe during the 18th and 19th centuries. In the mid-19th century, commodity futures markets, particularly for agricultural products, became more organised in response to the needs of farmers and merchants who sought to hedge against adverse price fluctuations.

The modern futures markets took a significant step forward with the establishment of the Chicago Board of Trade (CBOT) in 1848. The CBOT initially focused on agricultural commodities, particularly grains such as wheat and corn. It introduced standardised futures contracts to facilitate trading and reduce counterparty risk. The CBOT introduced standardised futures contracts in the 1860s and 1870s, specifying contract terms such as quantity, quality, and delivery conditions. This innovation enhanced market liquidity and efficiency.

The Chicago Mercantile Exchange (CME), where futures contracts for commodities like grain, livestock, and energy products are traded, was established in 1898 as a spin-off from the CBOT. The CME initially focused on agricultural commodities but expanded to include a broader range of products over time. The 1970s saw the introduction of financial futures contracts. The International Monetary Market (IMM), a division of the CME, launched currency futures in 1972, allowing market participants to hedge against currency risk. This marked the beginning of financial futures trading. Interest rate futures gained popularity in the 1970s and 1980s. The CBOT introduced Treasury bond futures in 1977, which provided a way for market participants to manage interest rate risk.

The Chicago Board of Options Exchange (CBOE) began trading options on common stocks in 1973 – becoming the first modern organised market for trading options. This success was soon mirrored to include expansion into options on fixed-income securities, currencies, stock- and bond-indices, as well as a variety of commodities.

The late 20th century saw the advent of electronic trading platforms, transforming the way financial instruments are bought and sold. Numerous exchanges transitioned from traditional open outcry systems to electronic platforms for greater efficiency. The late 20th century saw financial markets becoming increasingly globalised, with exchanges around the world increasingly interconnected. Cross-border trading and the globalisation of financial products has became more prevalent.

Very importantly, the latter part of the 20th century and the early 21st century saw the growth of derivatives markets and financial innovation. Exchanges introduced new financial instruments, including options (see, Chapter 4), swaps, and other derivatives (e.g., Chapter 16). In fact, the first organised market for trading options was the CBOE, which formally began trading options on stocks in 1973.

Regulatory oversight of financial markets exchanges increased over the years to ensure fair and transparent trading. Regulatory bodies, such as the U.S. Securities and Exchange Commission (SEC) and the Financial Conduct Authority (FCA), play a crucial role in overseeing exchanges. (See, Section 2.9.)

In summary, exchanges provide a platform for trading a wide range of financial instruments, from stocks, currencies, and bonds to commodities and derivatives. Consequently, exchanges play a vital role in the functioning of global financial systems.

1.3 EQUITY MARKETS

Mark Twain famously observed:

> October: This is one of the peculiarly dangerous months to speculate in stocks. The others are July, January, September, April, November, May, March, June, December, August, and February.

Private companies often desire to obtain capital to expand or continue their business operations. A company divides its ownership into shares or stocks – each share representing a portion of ownership in the company. Ownership represents a right to financial performance of the company and the strategic management thereof. Market capitalisation represents the total value of a company's outstanding shares of stock and is calculated by multiplying the stock price by the number of outstanding shares.

Public markets involve companies that have issued shares to the public, and these shares are traded on stock exchanges like the NYSE or NASDAQ, and these are traded by retail investors, mutual funds, pension funds, hedge funds, speculators, arbitrageurs, and exchange traded funds (ETFs) (see, Chapter 2). Private markets involve shares of companies that are not publicly traded and are typically bought and sold in private transactions – there has been substantial growth in so-called Private Equity markets in recent years.

Stock prices are determined by supply and demand in the market. Factors like company performance, economic conditions, and investor sentiment directly influence stock prices. Most companies distribute a portion of their profits to shareholders in the form of dividends. Dividends are typically paid out regularly, which provide an income to shareholders. Companies also do share buy-backs with their retained earnings from time to time.

Stock market indices, such as the S&P500, Dow Jones Industrial Average (DJIA), and FTSE 100, track the performance of a specific group of stocks and serve as benchmarks for the overall market. These indices are frequently followed by professional investors and they are, increasingly, the subject of passive investments through ETFs. There is also significant trading in derivatives on stock market indices. Options and futures on indices are widely used for hedging and speculative purposes as well as arbitrage activities, which frequently create significant liquidity.

Stock markets are regulated to ensure fair and transparent trading. Regulatory bodies, such as the SEC in the United States, oversee market activities. (See, Section 2.9.)

In summary, the equity market plays a crucial role in capital formation, which allows companies to raise funds by issuing shares to the public. It also provides investors with the opportunity to participate in the growth and success of companies (and countries) and to diversify their investment portfolios. However, investing in stocks carries risks, and the value of stocks can fluctuate based on various factors.

1.4 COMMODITY MARKETS

The commodity markets play a crucial role in the global economy, serving as a vital platform for the trading of raw materials and primary goods. The commodity markets are diverse and encompass a wide range of raw materials and primary goods. Understanding the characteristics of each segment of the commodities market is vital for market participants, as different factors and dynamics drive each market. This leads to unique risk management challenges as well as opportunities to gain from them.

Former trader, and editor of the *Gartman Letter*, Dennis Gartman provides some insight on commodities,

> Commodities are a unique asset class. They are the only asset class that is a tangible, consumable, and depleting resource.

Commodities markets can broadly speaking be categorised into two main types, namely physical markets and derivatives markets. The physical markets involve the actual exchange of goods, while derivatives markets facilitate the trading of financial instruments, which are derived from underlying commodities. The latter includes futures (typically physically settled) and various different options contracts. Derivatives contracts are essential for risk management and price discovery.

It is also very important to note that the rise of ETFs has led to the investor's? ability to access certain commodity markets directly instead of historically accessing these indirectly through equity-related investments. Examples of such ETFs include gold, platinum, palladium, rhodium, and oil.

Let us dig a bit deeper. Commodity markets can be classified into energy, metals, and agricultural products. We commence by looking at some energy-based commodity examples:

1. **Crude Oil:** One of the most traded commodities, crude oil is essential for the global energy industry. Futures and options on crude oil are traded primarily on the New York Mercantile Exchange (Nymex) and the Intercontinental Exchange (ICE).

    We frequently reference West Texas Intermediate (WTI) and Brent, which serve as two major benchmarks. In the US, WTI crude oil serves as the benchmark. Brent crude is now generally accepted to be the world benchmark as it is used to price a significant proportion of the world's internationally traded crude oil supplies.

2. **Natural Gas:** Used for heating, electricity generation, and as feedstock for various industries. Natural gas markets are influenced by supply, demand, and weather conditions.

3. **Lithium:** With the rise of electric vehicles and energy storage, lithium has gained prominence for its use in batteries.

4. **Carbon Credits:** Traded in carbon markets as part of efforts to reduce greenhouse gas emissions.

Secondly, let us consider some metals markets:

1. **Precious Metals:** Gold and silver are considered safe-haven assets and are often used for hedging against economic uncertainties. Gold is typically held by central banks, world-wide, as a store of value.

    Platinum, palladium, ruthenium, rhodium, and iridium are also part of this complex. (Iridium is one of the rarest metals on Earth, representing less than 1 part per billion of the Earth's crust.)

2. **Industrial Metals (Base Metals):** The industrial metals complex, also known as non-ferrous metals, comprises aluminium (corrosion-resistant and lightweight, aluminium is used in various industries,

including aerospace and packaging), copper (often referred to as 'Dr Copper' due to its sensitivity to economic trends), lead, nickel, tin, and zinc.

Copper, aluminium, zinc, and nickel are vital for construction, manufacturing, and infrastructure development. These metals are traded on several exchanges around the world but benchmark contracts are listed on the London Metals Exchange (LME). Prices are typically influenced by industrial demand and global economic conditions.

Thirdly, we consider agricultural commodity examples:

1. **Grains:** Wheat, corn, soybeans, and rice are staple foods and important components of global agricultural trade. Prices are heavily influenced by weather conditions, global demand, and government policies.

2. **Soft Commodities:** Coffee, cocoa, palm oil, rubber, sugar, and cotton fall into this category. Soft commodities are influenced by factors such as climate and changing consumer preferences. (For interest, trading in palm oil dates to as early as 3,000 BCE.)

3. **Livestock:** Cattle and hogs, as an example, are traded in livestock markets. Prices are affected by factors such as feed costs, disease outbreaks, and consumer demand for meat products.

Investor demand has more recently spurred the creation of several commodity indices. These typically represent a basket of commodities and are often used as benchmarks for commodity-related investments. Investors also look at using commodities to provide a hedge against inflation.

Several factors directly impact commodity prices, making the markets inherently volatile. Geopolitical events, supply and demand dynamics, weather conditions, and economic indicators all play a significant role. Political tensions in oil-producing regions, for example, can disrupt supply, which affects oil prices globally. Likewise, a poor harvest due to adverse weather conditions can lead to price spikes in agricultural commodities. Appropriately, Warren Buffett notes:

> In the commodities market, the rearview mirror is always clearer than the windshield.

Advancements in technology have led to transformation of the commodities markets. Electronic trading platforms have replaced traditional open outcry systems, thus facilitating faster and more efficient transactions. The use of algorithms and artificial intelligence for trading strategies and market analysis has become increasingly prevalent, which influences market liquidity and price discovery. The commodities markets are highly influenced by globalisation trends and the rise of emerging markets. As developing economies grow, their demand for commodities increases, and, thus, affects global supply and demand dynamics. Additionally, the interconnectedness of financial markets across borders amplifies the impact of events in one region on commodities prices world-wide.

The commodities markets will continue evolving in response to changing global dynamics. Technological advancements, increased emphasis on sustainability, and geopolitical developments will likely shape the future landscape. Additionally, as the world transitions towards cleaner energy sources, commodities related to renewable energy, such as lithium and rare earth metals, will feature prominently. Investors are increasingly placing emphasis on commodities with lower carbon footprints, and there is a rising demand for ethically sourced materials.

In summary, the commodity markets remain a complex and dynamic part of the global financial system. Understanding the interplay of various factors that influence prices, staying abreast of technological advancements, and being mindful of environmental and social considerations are crucial for market participants. As the markets continue to evolve, adaptability and a keen awareness of emerging trends will be essential for success in this ever-changing landscape.

Perhaps, the final word belongs to Jesse Livermore as referenced by [73]:

> In the long run commodity prices are governed but by one law – the economic law of demand and supply. The business of the trader in commodities is simply to get the facts about the demand and the supply, present and prospective. He does not indulge in guesses about a dozen of things as he does in stocks.

1.5 CURRENCY MARKETS

The currency market, also known as the foreign exchange market (or forex market), is the global marketplace for buying and selling currencies. The forex market facilitates the exchange of currencies between participants, which can include governments, central banks, financial institutions, corporations, and individual traders.

The main currency pairs traded include the US Dollar (USD) against other major currencies such as the Euro (EUR), Japanese Yen (JPY), British Pound (GBP), and Swiss Franc (CHF), among others. The USD is used as the reserve currency of the world and most international trades settle against USD. Louis Gave [49] explains this toppingly:

> A reserve currency is thus a bit like a computer operating system – it pays to use the one that everyone else is using, and the more people use one system, the less incentive there is to switch.

The forex market is decentralised and operates 24 hours a day, across major financial centres in different time zones. It is the largest and most liquid financial market in the world. Based on the 2022 BIS Triennial Survey, average daily trading volume in the forex market was estimated to be around $7.5 trillion in April 2022, for example.

The history of the currency market can be traced back to ancient times when merchants engaged in cross-border trade. However, the modern forex market, as we now know it, began to take shape in the early 1970s with the collapse of the Bretton-Woods system. Prior to this, the international monetary system was based on fixed exchange rates, which were tied to the value of gold. The financier, J.P. Morgan, famously noted:

> Gold is money, everything else is credit.

The shift to a system of floating exchange rates marked the beginning of the forex market's contemporary structure.

In recent decades, advancements in technology have played a significant role in the evolution of currency markets. Electronic trading platforms have made it easier for many participants (both retail and professional) to access the market, which leads to increased trading volumes and liquidity.

In summary, the forex market plays a crucial role in the global economy by facilitating international trade and investment, managing currency risk, and serving as a mechanism for price discovery. It is influenced by various factors, including economic indicators, geopolitical events, interest rates, and market sentiment.

1.6 RATES MARKETS

James Carville, advisor to former president Bill Clinton, famously said:

> I used to think that if there was reincarnation, I wanted to come backs as the President, the Pope, or as a 0.400 baseball hitter. But now I would like to come back as the bond market. You can intimidate everybody.

Rates markets, also known as interest rate markets, form a critical segment of the global financial system. These markets serve as the epicentre for the pricing, trading, and management of interest rates, influencing a wide array of financial instruments and impacting the broader economy.

Central banks, such as the Federal Reserve (FED) in the United States or the European Central Bank (ECB), play a pivotal role in rates markets by setting short-term policy rates, influencing overall interest rate levels, and implementing monetary policy. Changes in policy rates influence borrowing costs, spending, and investment, thereby affecting aggregate economic activity.

Rates markets host a diverse range of financial instruments, each catering to specific needs and preferences of both issuers and investors. Some prominent instruments include:

1. **Government Bonds:** Bonds are probably the most common or well-known debt instruments, representing a loan made by investors to issuers. They typically pay periodic interest and return the principal at maturity. Sovereign bonds, such as U.S. Treasuries (UST), German Bunds, and Japanese Government Bonds, serve as benchmark instruments in rates markets. They are considered low-risk assets and play a crucial role in determining the risk-free rate.

2. **Inflation-Linked Bonds:** The global Inflation-linked bonds (ILBs) market is still relatively young with strong growth since the early 2000s. The UK took the lead by issuing inflation-linked gilts in 1981. ILBs are designed to protect investors from loss of purchasing power, hence these bonds typically link coupons and principal repayment to the official consumer price index.

3. **Treasury Bills (T-Bills):** Treasury bills are short-dated (maturities typically ranging between four weeks to one year) debt securities, issued at a discount to par value, by the government as a mechanism

to raise funds. Considered to be risk-free investments, T-bills are frequently used by money-market funds and also pledged as collateral.

4. **Corporate Bonds:** Companies issue bonds to raise capital, and the interest rates on these bonds are influenced by prevailing rates in the markets. Corporate bonds provide investors with varying risk and return profiles based on the creditworthiness of the issuing entity.

5. **Notes:** Similar to bonds, notes are shorter-term debt instruments with maturities ranging from one to ten years.

6. **Commercial Paper:** Short-term unsecured promissory notes issued by corporations to meet immediate funding needs.

7. **Convertible Securities:** These are hybrid instruments that allow investors to convert their debt holdings into equity under specified conditions.

8. **Interest Rate Swaps:** These derivatives allow parties to exchange fixed-rate and floating-rate cash flows. Interest rate swaps are widely used for managing interest rate risk and adjusting the overall interest rate exposure of a portfolio.

9. **Futures and Options:** Traded on organised exchanges, interest rate futures and options provide a means for market participants to hedge or speculate on future interest rate movements.

In summary, it is difficult to over-emphasise the importance of rates markets in the global financial system; they influence a multitude of economic variables and financial instruments and serve as the basis for comparative asset pricing and other complex financial instruments.

1.7 DERIVATIVES MARKETS

Derivatives markets play a crucial role in modern financial systems, serving as instruments for risk management, speculation, and price discovery, and, in many markets, actually serve to define the market. Based on the most recent BIS statistics [5]:

> The notional value of outstanding OTC derivatives reached $715 trillion at end-June 2023, up 16% (or $97 trillion) since end-December 2022.

These financial instruments derive their value from an underlying asset, index, or rate, allowing market participants to hedge against uncertainties or speculate on future price movements. Market size is typically measured on the basis of their notional size or value of outstanding contracts – which can be quite substantial as seen above.

The derivatives market encompasses a wide array of instruments, including futures contracts, options, swaps, and forward contracts, each with its unique characteristics and applications. In short, also discussed in Chapter 4, we have the following types of basic derivatives contracts:

1. **Futures Contracts:** Futures contracts are standardised agreements between two parties to buy or sell an asset at a predetermined future date and price. As an example, in the commodities market, a farmer may use a futures contract to lock in the price of crops before the harvest, mitigating the risk of price fluctuations.

   These contracts are traded on organised financial exchanges, providing liquidity and facilitating transparency. The CME is a leading derivatives exchange where a variety of futures contracts are traded, covering commodities such as gold, oil, and agricultural products, as well as financial instruments like equity indices and interest rates. EUREX is a European derivatives exchange that offers a range of futures contracts, including those on equity indices, interest rates, and commodities. The LME is a key global metals exchange, where futures contracts for base metals such as copper, aluminium, and zinc are actively traded.

2. **Forwards:** Forward contracts are customised agreements between two parties to buy or sell an asset at a future date for a price agreed upon today. Unlike futures contracts, forwards are typically traded over-the-counter (OTC) and are not standardised. Importers and exporters often use forward contracts to hedge against currency fluctuations.

3. **Swaps:** Swaps involve the exchange of cash flows or other financial instruments between two parties over a specified period. Interest rate swaps, for example, allow entities to exchange fixed-rate and variable-rate interest payments. Companies may use interest rate swaps to manage interest rate risk on their debt. Significant parts of the interest rate swap market trades on an OTC basis.

The ICE is a global exchange that facilitates the trading of interest rate swaps. Participants can use ICE Swap Trade, an electronic platform, to execute standardised interest rate swap contracts. (In addition to interest rate swaps, ICE facilitates the trading of currency futures. Traders can access a variety of currency pairs through platforms like ICE Futures U.S.)

4. **Options:** Options grant the holder the right, but not the obligation, to buy (call option) or sell (put option) an asset at a specified price before or at expiration. Options are widely used for hedging and speculation. For instance, an investor might purchase put options on a reference stock index, such as the S&P500, to hedge against potential declines in the value of its stock portfolio.

    The CBOE is a major options exchange in the United States. It lists options contracts on various equity securities, as well as indexes, like the S&P500, and ETFs. Euronext is a pan-European exchange that operates options markets in several countries. The Amsterdam Options Market, for example, facilitates the trading of options on individual stocks and indices.

Don Chance [25] emphasises the importance of the modern derivatives markets thus:

> It has been said in the past that derivatives are kind of a side show, where the main event takes place in the money and capital markets. One could attend the side show without taking part in the main event and vice versa. With respect to derivative and money/capital markets, that is simply not true today. Derivatives are so widely used that even if one has no intention of using them, it is important to understand how they are used by others and what effects, positive and negative, they could have on money and capital markets.

Let us list some roles of the modern derivatives markets:

1. **Risk Management:** Derivatives markets enable participants to manage risks such as price, interest rate, currency, and commodity risks. For example, airlines may use futures contracts to hedge against fluctuations in fuel prices to aid in stabilising their operational costs.

2. **Price Discovery:** Derivatives markets contribute to price discovery by reflecting market expectations and sentiment. The prices of derivatives are influenced by factors such as supply and demand dynamics, economic indicators, and geopolitical events. Traders and investors use this information to make informed decisions about the future direction of asset prices.

3. **Liquidity and Efficiency:** Derivatives markets enhance overall market liquidity and efficiency. By providing standardised instruments that can be easily bought or sold, derivatives contribute to the smooth functioning of financial markets. This liquidity benefits both hedgers and speculators.

4. **Financial Innovation:** Derivatives markets drive financial innovation by offering new instruments and strategies. Exotic options, structured products, and other complex derivatives have been developed to address specific risk-management needs or to create investment opportunities.

In summary, derivatives markets have become integral to the functioning of modern financial systems, offering participants tools to manage risk and enhance market efficiency. While derivatives provide opportunities for speculation, their primary role lies in risk mitigation, aiding businesses, investors, and financial institutions in navigating an increasingly complex and dynamic economic landscape. As these markets continue to evolve, it is essential to strike a balance between innovation and risk management to maintain the stability and integrity of the financial system.

The final word here belongs to Nassim Nicholas Taleb [100], who emphasises the importance of risk management:

> Every derivative user or trader needs to be able to account for the effect of either the passage of time or the movement of the underlying assets on the portfolio. Issues of pricing are therefore overstated. Most money is made or lost because of market movement, not because of mispricing. Often the cause is mishedging. Most commonly, losses result from a poor understanding of liquidity and the shape of the statistical distribution.

CHAPTER 2

Market Players

The intelligent investor is a realist who sells to optimists and buys from pessimists.

— BENJAMIN GRAHAM (1894–1976)

I've seen a lot more go to zero than infinity.

— JAMES CHANOS

There is time to go long, time to go short and time to go fishing.

— JESSE LAURISTON LIVERMORE (1877–1940)

The goal of a successful trader is to make the best trades. Money is secondary.

— ALEXANDER ELDER

After retiring from the military, my dad worked in the real estate industry. He once told me: "The deal of the century comes along every year."

— BILL MCBRIDE

2.1 INTRODUCTION

Financial markets evolve and change over time. The players in these markets become more sophisticated over time and have access to newer technologies but frequently also change their focus. Mauboussin [78] writes:

> How do we know that money managers are increasingly short-term oriented? We see it in the portfolio turnover data. Average portfolio turnover has surged in recent decades, going from roughly 20 percent in the 1950s to well nearly 100 percent today.

There are numerous participants in every market. These participants can also take on dual roles, for example, speculation and investing; however, in what follows, we provide a short overview of the main activities. We need to understand the different players in order to gain a fuller understanding of the financial markets and our desire to create and hedge products.

We shall differentiate between speculators, investors, arbitrageurs, and hedgers. We also consider market-makers and hedge funds. We also ponder the role of central banks. While regulators are not directly involved in the markets, they do influence trading activity; thus, we briefly discuss them as well.

2.2 SPECULATORS

Speculation in the market is the subject of volumes of books. Mark Twain famously wrote:

> There are two times in a man's life when he should not speculate: when he can't afford it, and when he can.

In the book *Reminiscences of a Stock Operator* [73], Jesse Livermore recalls this delightful discussion:

> One day my friend came to me and asked me,
> "Have you covered?"
> "Why should I?" I said.
> "For the best reason in the world."
> "What reason is that?"
> "To make money."

Speculators in financial markets try to make money by buying and selling securities, usually within a short timeframe. They usually operate on limited amounts of capital and employ significant amounts of leverage. In principle, any goods any goods, or financial instrument, could be the subject of speculative buying/selling; however, if the goods/instrument is illiquid, speculative activity will lead to higher costs (or slippage) and becomes unattractive.

Jesse Livermore said:

> Speculation is hard and trying business, and a speculator must be on the job all the time or he'll soon have no job to be on.

Speculators seek to profit from price movements – they frequently follow some technical analysis of price patterns or seek to exploit short-term price reactions to news events. Successful speculators follow a highly disciplined approach to their trading with emphasis on curtailing losses. Again, wise advice from Jesse Livermore [73]:

> Getting angry doesn't get a man anywhere. More than once it has been borne in on me that a speculator who loses his temper is a goner.

Speculators are important in financial markets as they provide a source of liquidity and are vital for price discovery.

2.3 INVESTORS

The legendary economist, John Maynard Keynes, described the difference between investors and speculators thus:

> Investing is an activity of forecasting the yield over the life of the asset; speculation is the activity of forecasting the psychology of the market.

Warren Buffett is probably world-wide synonymous with long-term stock market investments. He has famously written:

> Our favourite holding period is forever.

He is also known for saying:

> I never attempt to make money on the stock market. I buy on the assumption that they could close the market the next day and not reopen it for five years.

We therefore typically associate investments with a longer-term view. Long-term investors form a critical pillar of the financial markets; they are typically characterised by their patient and strategic approach to ultimate wealth creation.

Unlike short-term traders (i.e., speculators) with limited capital, who seek quick gains, long-term investors focus on sustained growth over extended periods, while exhibiting sustained patience and discipline. While short-term traders may be susceptible to volatility, long-term investors are better positioned to ride out market downturns and benefit from the overall upward trajectory of the market. They understand that market fluctuations are inevitable but choose to weather short-term volatility in pursuit of their overarching financial investment goals.

Paul Samuelson has a succinct view on investments:

> Investing should be more like watching paint dry or watching grass grow. If you want excitement, take $800 and go to Las Vegas.

Long-term investors carefully research and select assets based on their growth potential, financial health, risk characteristics, and alignment with broader economic trends. Their strategic foresight enables them to endure through market cycles and capitalise on long-term trends. Fundamental analysis is a definite cornerstone of long-term investment strategies. Investors assess the underlying financial health and performance of companies, ensuring that their chosen investments have solid foundations for sustained growth.

Long-term investors prioritise wealth accumulation over time. By remaining invested for extended periods, they benefit from the compounding of returns, allowing their assets to grow exponentially. The long-term horizon enables investors to weather short-term market fluctuations. Investors make money from the longer-term trend in markets; this pertinent summary is given by Steve Leuthold:

> History teaches us that investors behave wisely ... once they have exhausted all other alternatives.

Long-term investors contribute to the stability of financial markets by providing a steady and stable source of capital. Their commitment to holding investments through market fluctuations dampens short-term volatility, which contributes to overall market stability. As custodians of capital, long-term investors play a crucial role in shaping the trajectory of businesses, industries, and societies, which fosters a resilient and prosperous financial landscape. The investor's ability to withstand short-term volatility while focusing on the long game exemplifies the strategic prowess that underpins their enduring impact.

2.4 HEDGERS

A very useful general definition of hedging is given by Gregory Connor:

> Hedging is the purchasing of an asset or portfolio of assets in order to insure against wealth fluctuations from other sources.

Investment sage Howard Marks said the following about hedging:

> Hedging is not about making money but about protecting what you have.

Hedgers in financial markets are individuals or entities that use financial instruments to reduce or mitigate the risk associated with price fluctuations in an underlying asset. These participants engage in hedging strategies to protect themselves from potential adverse movements in the market that could negatively impact their financial positions.

The primary goal of hedging, therefore, is to offset potential losses in one investment by taking an opposing position in another investment or financial instrument (or combinations thereof). By doing so, hedgers aim to stabilise or reduce their overall portfolio exposure against market uncertainties. There are various types of hedgers, and they can be found across different financial markets, including commodities, currencies, and financial derivatives.

Here are a few examples of hedgers and hedging activity:

1. **Commodity Producers:** Companies involved in the production of commodities, for example, farmers or mining companies, may use hedging to protect themselves from fluctuations in commodity prices. A farmer might use futures contracts to lock in a selling price for their crops before the crops are harvested – this might be a financing requirement as well.

2. **Importers and Exporters:** Businesses frequently engage in international trade, hence they face currency risk due to fluctuations in exchange rates. Importers and exporters can use currency hedging tools, such as forward contracts or various option strategies, to protect themselves from adverse currency movements.

3. **Investment Funds:** Mutual funds, hedge funds, and other investment vehicles may use hedging strategies to manage risks and exposures associated with their portfolios. These funds will typically use combinations of options and futures contracts to hedge against potential declines in the value of their risky asset holdings.

4. **Individual Investors:** Individual investors may also employ hedging strategies to protect their investment portfolios from adverse market movements. This could involve using options contracts or other derivatives to offset potential losses.

5. **Interest Rate Hedgers:** Entities with exposure to interest rate risk, such as financial institutions or companies with variable-rate debt, may use interest rate swaps or futures contracts to hedge against interest rate fluctuations.

It is important to understand that hedgers could work from the long or short side – an insurer hedging an equity-linked investment for a client could hold equities to match its liability while shorting bonds to protect against rate increases.

2.5 ARBITRAGEURS

Arbitrage, a fundamental concept in financial markets, refers to the practice of exploiting price differentials of identical or similar financial instruments across various markets to generate profits with minimal risk. This activity is rooted in the efficient market hypothesis, which posits that asset prices reflect all available information, leaving no room for profitable trading strategies. However, arbitrageurs challenge this notion by identifying and capitalising on temporary market inefficiencies. Mathematician and market entrepreneur, Edward Thorp, wisely cautions:

> There are inefficiencies in the market, but they're not easy to demonstrate, and I think that needs to be done before one shifts money in that direction.

Arbitrage opportunities arise in various forms, and market participants employ different strategies to capitalise on them. Three common types of arbitrage include:

1. **Spatial Arbitrage:** Spatial arbitrage involves exploiting price differentials of the same asset in different geographic locations. With globalisation and advances in technology, traders can quickly identify and execute trades to benefit from price imbalances between markets.

2. **Temporal Arbitrage:** Temporal arbitrage focuses on exploiting price differences that occur over time. This can involve taking advantage of pricing anomalies resulting from delays in information dissemination, market reactions to news, or variations in trading hours between different markets.

3. **Statistical Arbitrage:** Statistical arbitrage relies on quantitative models and statistical analysis to identify mispriced assets. Traders using this strategy often analyse historical price data, correlations between assets, and other quantitative factors to make informed investment decisions.

Arbitrageurs employ a variety of methods to execute their strategies efficiently. These methods include manual trading, algorithmic trading, and high-frequency trading. Algorithmic and high-frequency trading have gained prominence due to their ability to execute trades at speeds beyond human capability, which allows arbitrageurs to capitalise on fleeting opportunities in the market.

While these so-called arbitrage activities can yield profits, they are not without risks. Michael Lewis [68] captured this entertainingly:

In need of a euphemism for what we did with other people's money, we called it 'arbitrage', which was just plain obfuscation.

Several factors contribute to the risks associated with arbitrage:

1. **Execution Risk:** The speed at which arbitrage opportunities arise and disappear introduces execution risk. Delayed order execution can erode potential profits or even result in losses.

2. **Market Risk:** Fluctuations in market conditions can impact the success of arbitrage strategies. Changes in interest rates, geopolitical events, or sudden market shocks can lead to unexpected price movements, affecting the profitability of arbitrage trades.

3. **Liquidity Risk:** Arbitrage opportunities are often short-lived, and illiquid markets may pose challenges when entering or exiting positions. This liquidity risk can result in slippage, where the executed price differs from the expected price.

Arbitrage activities contribute to market efficiency by narrowing price disparities and ensuring that prices reflect all available information. As arbitrageurs exploit mispricings, they facilitate a more efficient allocation of resources in the market. However, the increasing prevalence of algorithmic and high-frequency trading has sparked debates about their impact on market stability and the potential for excessive speculation.

Arbitrage activities play a crucial role in financial markets by contributing to price discovery and market efficiency. While providing opportunities for profit, arbitrageurs must navigate various risks, including execution, market, and liquidity risks. The evolution of technology has transformed the landscape of arbitrage, with algorithmic and high-frequency trading becoming dominant methods. As financial markets continue to evolve, understanding and managing the risks associated with arbitrage activities will be essential for market participants seeking to capitalise on price differentials while maintaining a balanced and sustainable approach to trading.

Perhaps, the final word on arbitrage belongs to Mauboussin [78]. He writes:

> Further, arbitrage is insufficient to bring markets back to efficiency. So inefficiency is the rule, and efficiency the exception.

Active portfolio management in a fundamentally inefficient market is a mug's game.

We shall focus on the demands of no-arbitrage in the context of pricing derivatives and for establishing pricing bounds. See, Chapter 6, where we show that the concept of no-arbitrage forms a crucial cornerstone of the pricing of derivative securities.

2.6 MARKET MAKERS

Market makers play a crucial role in financial markets by facilitating the buying and selling of financial instruments such as stocks, bonds, and derivatives. Their primary function is to provide liquidity to the market, ensuring that there is a smooth and efficient flow of trades.

While providing liquidity and maintaining market efficiency are key functions, market makers are also motivated by profit. They earn a profit from the bid-ask spread, that is, the difference between buying and selling prices. Effective risk management and accurate pricing are essential for their profitability. Here are key aspects of the role of market makers:

1. **Liquidity Provision:** Market makers provide continuous buy and sell prices for a specific set of financial instruments. By doing so, they stand ready to buy or sell these instruments, which ensures that there is a readily available market for traders and investors.

2. **Price Stabilisation:** Market makers help in maintaining price stability by narrowing the bid-ask spread – the difference between the highest price a buyer is willing to pay (bid) and the lowest price a seller is willing to accept (ask). A narrower spread reduces trading costs for market participants and fosters a more liquid market.

3. **Risk Management:** Apart from execution risk, market makers take on numerous risks in their role. They need to manage their inventory of financial instruments effectively to ensure they can meet their obligations to buy or sell at quoted prices. This involves monitoring market conditions, adjusting their bid-ask spreads, and hedging their positions.

4. **Facilitating Trading:** Market makers facilitate the trading process by providing a counterparty for buyers and sellers. In the absence of

market makers, finding a willing buyer or seller at any given time might be challenging, particularly for less-liquid securities.

5. **Regulatory Compliance:** Market makers operate within a regulatory framework that typically varies by jurisdiction. They must adhere to rules and regulations that govern their activities, ensuring fair and transparent markets. Compliance with these regulations helps maintain market integrity.

6. **Technology and Automation:** In many modern financial markets, market making is often highly automated. Market makers use advanced trading algorithms and technology to respond to market changes and execute trades efficiently. This automation allows for faster and more accurate pricing.

Overall, market makers play a critical role in ensuring the smooth functioning of financial markets, contributing to liquidity, price stability, and efficient trading.

2.7 HEDGE FUNDS

A hedge fund is a pooled investment fund that employs a variety of strategies to generate returns for its investors. These funds are typically open to accredited investors and have the flexibility to invest in a wide array of financial instruments, including stocks, bonds, derivatives, currencies, and commodities. Unlike mutual funds (typically long-only investments), hedge funds are subject to less regulatory oversight, allowing fund managers greater flexibility in their investment decisions.

Nouriel Roubini states how controversial investments in hedge funds and their associated management fees, can be:

> Hedge funds are like the rock stars of finance. They command huge fees, have legions of devoted fans, and often burn out in a blaze of glory.

Importantly, though, on a philosophical level, hedge funds are characterised by their pursuit of absolute returns, meaning they aim to generate positive returns regardless of the broader market conditions. This is in contrast to traditional investment funds, which often benchmark their performance against a market index.

The concept of hedge funds can be traced back to the mid-20th century, with the establishment of the first hedge fund by Alfred Winslow Jones in 1949. Jones's fund used a long-short equity strategy, allowing it to profit from both rising and falling markets. Over the decades, hedge funds evolved, and their strategies diversified to include global macro, event-driven, quantitative, and arbitrage strategies, among numerous others.

1. **Long-Short Equity:** This strategy involves taking both long and short positions in stocks, aiming to profit from both rising and falling stock prices.

2. **Global Macro:** Global macro funds focus on macroeconomic trends and make investments based on their expectations for changes in interest rates, currencies, and commodity prices.

3. **Event-Driven:** Event-driven funds capitalise on corporate events such as mergers, acquisitions, bankruptcies, and other significant developments that can impact on a company's stock price.

4. **Quantitative Strategies:** Quantitative hedge funds use mathematical models and algorithms to identify investment opportunities, relying on data analysis and statistical techniques.

5. **Arbitrage:** Arbitrage strategies involve exploiting price differentials between related assets or markets, seeking to capture profits from inefficiencies.

Hedge funds contribute to market liquidity by actively participating in buying and selling financial instruments. This liquidity can enhance the efficiency of financial markets. Hedge funds often act as risk managers by employing strategies that seek to mitigate risks and reduce portfolio volatility. This risk management function can have a stabilising effect on markets.

Hedge funds provide investors with alternative investment opportunities and contribute to market efficiency. Their ability to navigate diverse strategies allows them to adapt to changing market conditions. However, the industry also faces challenges, and ongoing discussions regarding regulatory frameworks and investor protections underscore the importance of striking a balance between innovation and risk management in the dynamic world of hedge funds.

2.8 CENTRAL BANKS

No discussion on players in the financial markets could be complete without noting the proverbial 'the elephant in the room', namely central banks. In the words of Richard Layard, as referenced in [10],

> Central banks have never been more powerful than now. Monetary policy has become the central tool of macroeconomic stabilization, and in more and more countries monetary policy is in the hands of an independent central bank.

Central banks serve as the cornerstone of modern monetary systems. They wield significant influence over economic stability and growth. Through their multifaceted functions and roles, which we will briefly discuss below, they strive to maintain price stability, promote financial stability, and support sustainable economic development. As guardians of monetary stability, central banks play a vital role in shaping the trajectory of national and global economies.

It is instructive to ponder the words of Alan Blinder, a former Vice-Chairman of the Board of Governors of the Federal Reserve Board, on central banks [10]:

> Monetary policymakers have certain objectives – such as low inflation, output stability, and perhaps external balance – and certain instruments to be deployed in meeting their responsibilities, such as bank reserves or short-term interest rates. Unless it has a single goal, the central-bank is forced to strike a balance among competing objectives, that is, to face up to various *trade-offs*.

Central banks play a crucial role in ensuring the smooth functioning of the economy and fostering confidence in the financial system. Their significance stems from:

1. **Independence:** Central banks are often granted a high degree of independence from political interference. This allows them to pursue monetary policy objectives without undue political influence. This independence enhances credibility and effectiveness in achieving macroeconomic goals.

2. **Credibility and Trust:** Central banks' commitment to price stability, and sound monetary policy, instills confidence in investors, consumers, and businesses. Their transparent communication and consistent policy actions contribute to building trust in the financial system.

3. **Global Impact:** The decisions and actions of major central banks have far-reaching implications beyond their national borders. Changes in interest rates or monetary policy stance by influential central banks can affect global financial markets, exchange rates, and capital flows.

Central banks play various roles in the economy, strategically serving as:

1. **Guardians of Price Stability:** One of the primary objectives of central banks is to maintain price stability by controlling inflation. By adjusting interest rates and influencing the money supply, central banks strive to keep inflation within a target range conducive to economic stability and growth.

2. **Lenders of Last Resort:** During financial crises or liquidity crunches, central banks act as lenders of last resort, providing emergency funding to financial institutions to prevent systemic collapse. This role helps restore confidence in the financial system and averts widespread economic turmoil.

3. **Economic Stabilisers:** Central banks use monetary policy tools to counteract fluctuations in economic activity, such as recessionary downturns or overheating expansions. By adjusting interest rates and liquidity conditions, they aim to stabilise output and employment levels, fostering sustainable economic growth over the long term.

4. **Promoters of Financial Stability:** Through regulatory oversight and supervision, central banks promote the stability and resilience of the financial system. By identifying, and mitigating, systemic risks, they reduce the likelihood of financial crises and enhance the overall health of the banking sector.

Central banks also typically perform several key functions, each of which have an impact on financial markets:

1. **Monetary Policy:** Perhaps the most prominent function of central banks is formulating and implementing monetary policy. Through mechanisms such as interest rate adjustments, open market operations, and reserve requirements, central banks aim to achieve macroeconomic objectives like price stability, full employment, and sustainable economic growth.

2. **Currency Issuance and Management:** Central banks have the sole authority to issue and manage a nation's currency. They regulate the money supply to ensure stability and prevent excessive inflation or deflation.

3. **Banker to the Government:** Central banks often act as bankers to their respective governments, managing their accounts, facilitating borrowing through bond issuance, and providing financial services. This role helps in fiscal management and ensures smooth functioning of the government's financial operations.

4. **Regulator and Supervisor of Financial Institutions:** Central banks are responsible for supervising and regulating commercial banks and other financial institutions within their jurisdiction. They establish prudential regulations to safeguard the stability of the financial system and protect depositors' interests.

5. **Custodian of Foreign Exchange Reserves:** Central banks manage a nation's foreign exchange reserves, which are crucial for maintaining exchange rate stability and meeting international financial obligations.

Former president of the ECB, Mario Draghi noted:

> Central banks are the guardians of price stability, charged with the critical mission of preserving the value of our money.

Indeed, Christine Lagarde, president of the ECB declared:

> Central banks are the gatekeepers of our modern financial system, wielding immense power over the economy and markets.

Noting central bank influences on the financial markets, Mohammed Erian quipped:

Central banks are like the Wizard of Oz, pulling levers behind a curtain. When they speak, markets listen.

Let us, therefore, look at a number of previously unconventional policy actions that central banks have taken in recent years. These actions have had significant effects on economies and trends in financial markets.

1. **Quantitative Easing:** Quantitative easing (QE) is a monetary policy tool employed by central banks to stimulate the economy, primarily when traditional methods, such as adjusting interest rates, are insufficient. It involves the central bank purchasing long-term securities, typically government bonds or mortgage-backed securities, from the open market to increase the money supply and lower long-term interest rates. These purchases increase the reserves held by banks, providing them with more liquidity to lend to businesses and consumers. Consequently, this boosts spending, investment, and economic activity. Consider the following examples:

    a. **United States (2008–2014):** In response to the global financial crisis of 2007–2008, the U.S. Federal Reserve initiated multiple rounds of QE. The FED purchased trillions of dollars' worth of long-term Treasury securities and mortgage-backed securities. These actions aimed to stabilise financial markets, lower borrowing costs, and support economic recovery.

    b. **European Union (2015–2018):** The ECB implemented a large-scale QE programme to combat stagnant growth and deflationary pressures in the euro zone. The ECB purchased government bonds and other assets, totalling over €2 trillion, to boost liquidity and stimulate lending in the region.

    c. **Japan (2013–2024):** The Bank of Japan (BoJ) has pursued an aggressive QE policy as part of its broader efforts to overcome decades of economic stagnation and deflation, commonly referred to as Japan's 'Lost Decade.' The BoJ has purchased vast amounts of government bonds and ETFs to lower borrowing costs, spur inflation, and revitalise economic growth.

    d. **COVID-19 Pandemic Support:** Various central banks, including the US Federal Reserve, ECB, Bank of England (BOE), and others, significantly expanded their asset purchase programmes to mitigate the world-wide economic fallout from the COVID-19 pandemic. These programmes entailed purchasing a wide range of assets, including government bonds, mortgage-backed securities, corporate bonds, and even ETFs with the aim of injecting liquidity into financial markets, providing support for credit markets, and lowering of borrowing costs.

2. **NIRP:** The ECB implemented a negative interest rate policy (NIRP) as part of its monetary policy toolkit. The NIRP is a non-conventional monetary policy tool where central banks set key interest rates below zero.

    The primary objective of NIRP is to stimulate borrowing, lending, and spending in the economy by penalising commercial banks for holding excess reserves and incentivising them to lend money to businesses and consumers. The intention behind NIRP is to transmit the negative rates throughout the financial system, ultimately leading to lower borrowing costs for businesses and consumers. As commercial banks face higher costs for holding excess reserves, they are more likely to reduce deposit rates for their customers or even pass on negative rates to depositors. This can encourage spending and investment, as the cost of borrowing decreases.

    Negative interest rates influence currency exchange rates and financial markets. In theory, a negative interest rate differential between one currency and another can weaken the currency with the negative rates, as investors seek higher returns elsewhere. Additionally, the NIRP can lead to distortions in financial markets, as investors may search for higher yields in riskier assets or asset bubbles may form due to the search for returns.

3. **Forward Guidance:** The so-called forward guidance is a communication strategy employed by central banks to provide guidance to financial markets, businesses, and the public on the likely future path of monetary policy.

    The tool is used to manage expectations and influence financial conditions without necessarily making immediate changes to interest rates or other monetary policy tools. During the COVID-19 pandemic, for example, central banks provided guidance regarding their commitment to provide liquidity and maintenance of QE programmes and, thus, reassured reassuring investors.

The interaction between financial markets and central bank actions is clearly of material importance. Perhaps, the final words belong to Alan Blinder [10]:

> Now, in a literal sense, independence from the financial markets is both unattainable and undesirable. Monetary policy works *through* markets, so *perceptions* of likely market reactions must be relevant to policy formulation and *actual* market reactions must be relevant to the timing and magnitude of policy effects. There is no escaping this. It's important and of consuming interest to practical central bankers.

He cautions [10]:

> My point is simply that delivering the policies that markets expect – or indeed demand – may lead to very poor policy.
>
> The danger is greater now than ever, I believe, because the current-prevailing view of financial markets among central bankers is one of deep respect. The broad, deep, fluid markets are seen as repositories of enormous power and wisdom. In my personal view, the power is beyond dispute, but the wisdom is somewhat suspect.

2.9 REGULATORS

Christine Lagarde, now the president of the ECB, remarked on regulation:

> Markets without rules can become a jungle. Proper regulation is the compass that keeps us on the right path.

Regulators of the financial system play a crucial role in maintaining the stability, integrity, and fairness of financial markets. Their responsibilities include overseeing financial institutions, enforcing rules and regulations, and protecting the interests of investors and the broader economy. Here are key aspects of the role of regulators in the financial system:

1. **Market Integrity:** Regulators work to ensure the integrity of financial markets by preventing fraudulent activities, market manipulation, and other forms of misconduct. They establish and enforce rules that

promote fair and transparent trading practices, fostering confidence among market participants.

2. **Financial Stability:** Regulators monitor and assess the overall stability of the financial system. They implement measures to prevent systemic risks, such as excessive leverage, and address issues that could lead to financial crises. This involves supervising financial institutions to ensure they have sound risk management practices.

3. **Investor Protection:** One of the primary roles of financial regulators is to protect the interests of investors. They set rules that require disclosure of relevant information to investors, ensure fair and equal treatment, and establish mechanisms for dispute resolution. Investor protection measures contribute to maintaining trust in the financial markets.

4. **Prudential Regulation:** Regulators implement prudential regulations to ensure the safety and soundness of financial institutions. This includes setting capital adequacy requirements, liquidity standards, and other measures to mitigate the risk of financial institutions failing or causing broader systemic issues.

5. **Market Surveillance:** Regulators engage in continuous monitoring and surveillance of financial markets to detect and address any irregularities or market abuses. They may use technology and data analysis tools to identify unusual trading patterns, market manipulation, or other potential threats to market integrity.

6. **Enforcement of Laws and Regulations:** Regulators have the authority to enforce laws and regulations governing the financial sector. This authority allows regulators to conduct investigations, impose sanctions, and take legal action against individuals or institutions that violate the established rules. Enforcement actions serve as deterrents and help maintain the rule of law.

7. **Policy Development:** Regulators actively participate in the development and refinement of financial policies. They collaborate with other regulatory bodies, central banks, and international organisations to address emerging challenges, adapt regulations to changing market conditions, and enhance the effectiveness of the regulatory framework.

8. **Risk Assessment and Management:** Regulators assess various risks within the financial system, including credit risk, market risk, and operational risk. They work to ensure that financial institutions have robust risk management practices in place and can withstand adverse economic conditions.

9. **Consumer Protection:** Regulators are responsible for protecting consumers in financial transactions. This includes ensuring that financial products and services are transparent and fair, and that consumers are provided with adequate information to make informed decisions.

10. **International Coordination:** In an increasingly interconnected global financial system, regulators often collaborate on an international level to address cross-border challenges, harmonise regulatory standards, and enhance the consistency of regulatory approaches across jurisdictions.

Regulatory frameworks, therefore, play a crucial role in ensuring fair and transparent trading in markets. Regulatory bodies oversee exchanges, monitor market participants, and implement measures to prevent market manipulation and fraud. Changes in regulations, such as position limits and reporting requirements, can have a profound impact on market behaviour.

Cassidy [24] writes, with reference to regulations:

> Minsky advanced the view that free market capitalsim is inherently unstable, and that the primary source of the instability is the irresponsible actions of bankers, traders, and other financial types. Should the government fail to regulate the financial sector effectively, Minsky warned, it would be subject to periodic blowups, some of which could plunge the entire economy into lengthy recessions.

Overall, the role of regulators in the financial system is multifaceted, encompassing the promotion of market integrity, investor protection, financial stability, and the effective functioning of financial markets. Their actions are essential for maintaining public trust and confidence in the financial system and their effects on markets (and market-dynamics) should never be underestimated.

CHAPTER 3

Rates

Would you rather have one dollar today or one dollar in one year's time?

– STEPHEN BLYTH [11]

Interest rates are the reward for deferring consumption.

– PAUL A. SAMUELSON (1915–2009)

Interest rates are the cost of money, and their movements can have a profound impact on your life.

– SUZE ORMAN

Interest rates are the thermometer of the economy.

– ALAN GREENSPAN

Interest rates are to asset prices what gravity is to the apple.

– RAY DALIO

3.1 INTRODUCTION

Interest is the cost of money. Consider the following definition by Ingersoll:

> Interest is payment for use of funds over a period of time, and the amount of interest paid per unit of time as a fraction of the balance is called the interest rate.

Interest rates, and the cost of money, by implication, are at the centre of pricing financial derivatives. As an example, if we are trying to hedge a contingent claim, we need to factor the cost of carry for holding hedges into our analysis. We provide a précis below.

3.2 RATES

Using the above definition of interest, let us assume we invest an amount, N, (also referred to as the notional, principal amount or face value) at an annual rate, r (we compound interest annually to start). After a period of T years, we have that our investment has grown to

$$N(1+r)^T.$$

We could, of course, have more regular compounding of interest. So, in general, assuming we invest N at a rate, r, which is compounded m times per annum, we find our investment has grown to

$$N(1+r/m)^{mT}.$$

We frequently use the so-called continuously compounded rate, which will be attained in the limit as $m \to \infty$. Thus, our investment has grown to

$$Ne^{rT}.$$

The compounding frequency is normally evident from the specific instrument chosen; although, we specify rates conventionally on a per-annum basis.

As an example, a money market account represents the accumulated value of an investment at a rate, typically, compounded daily. This is typical of a cash-like investment.

3.3 BONDS

We could also consider bonds. Let us look at a basic concept, namely a zero-coupon bond (ZCB). In the interest of keeping matters easy, we ignore default risk. We buy the ZCB at time, t, and receive 1, for sure, at its maturity, T. We denote the value of the ZCB by $B(t, T)$; hence, $B(T, T) = 1$.

So, given the price of the ZCB, we can define the continuously compounded rate, $R(t,T)$, we receive on our investment, that is

$$B(t,T) = e^{-R(t,T)(T-t)}. \tag{3.1}$$

In other words, the spot rate, $R(t,T)$, can be thought of as an investment in the ZCB with maturity, T, accumulating at an average rate of $R(t,T)$ over the full period.

The instantaneous spot rate of interest can, therefore, now be defined as

$$r(t) = \lim_{T \downarrow t} R(t,T).$$

In the absence of uncertainty, and to avoid arbitrage, an investor needs to be in the same position by buying a ZCB, $B(t,T)$, and therefore guaranteeing an investment rate, $R(t,T)$, or to make subsequent roll-over investments of shorter bonds. Therefore, we need to have that

$$R(t,T) = \frac{1}{T-t} \int_t^T r(s)ds.$$

In essence, that means the continuously compounded long rate must be the average of the short rates. Equivalently, we have that

$$B(t,T) = \exp\left[-\int_t^T r(s)ds\right].$$

This allows us to define the money market account on a similar basis as:

$$M(t,T) = \exp\left[\int_t^T r(s)ds\right]. \tag{3.2}$$

In the special case where $r(\cdot)$ is constant, we have that

$$B(t,T) = e^{-r(T-t)},$$

and

$$M(t,T) = e^{r(T-t)}.$$

3.4 PRACTICAL ISSUES

The natural question, now, is where do we get information on $R(t,T)$? We will frequently refer to $R(t,T)$ as the spot yield curve. Parsimonious models, which are frequently used by central banks, governments, and many academic studies, invoke the parametric Nelson–Siegel form of the yield curve (see, [77], for example). This entails using a functional form of the yield curve as given by:

$$R(t,T) = \beta_0 + \beta_1 \left(\frac{1 - \exp(-\lambda(T-t))}{\lambda(T-t)} \right) + \beta_2 \left(\frac{1 - \exp(-\lambda(T-t))}{\lambda(T-t)} - \exp(-\lambda(T-t)) \right). \tag{3.3}$$

The parameters β_0, β_1, and β_2 can be interpreted in terms of the so-called level, slope, and convexity of the yield curve, corresponding, in principle, to the first three principal components of a yield curve decomposition. The λ parameter describes decay.

Fitting the parameters for Equation (3.3) is, in principle, a regression problem. In practice, the procedure of constructing yield curves from market data is, in general, based on two types of methods. Firstly, a 'best-fit' model, where a functional form, such as the Nelson–Siegel functional, is assumed and then parameters are calibrated to the market on a best-fit basis. The alternative method is an exact-fit methodology where we pre-select market instruments that are exactly repriced using bootstrapping, and then we would do some form of interpolation between the calculated rates.

3.5 FURTHER READING

A full discussion of interest rates is beyond the scope of this book. The intention here is to derive a basic working understanding of rate and basic bonds (ZCBs). There are numerous great books on interest rate models and interest rate derivatives. We mention Blyth [11], Björk [19], Cairns [20], Martellini et al. [77], and Wilmott et al. [105]. The book by Chancellor [27] provides an interesting history of rates.

CHAPTER 4

Derivatives

What makes derivatives important is not so much the size of the activity, as the role it plays in fostering new ways to understand, measure, and manage financial risk. Through derivatives, the complex risks that are bound together in traditional instruments can be teased apart and managed independently, and often more efficiently.

– GLOBAL DERIVATIVES STUDY GROUP - 1993

Derivatives have permitted the unbundling of financial risks.

– ALAN GREENSPAN

In our view, however, derivatives are financial weapons of mass destruction ...

– WARREN BUFFET

Derivatives typically involve little up-front payment and are a contract between two parties to exchange a flow of returns or commodities in the future. The principle of derivatives instruments is simple, but if you want to make it complicated there are many lawyers, and investment bankers who will help you – at a (significant) price.

– MERVYN KING [64]

4.1 INTRODUCTION

What are financial derivatives? Nassim Nicholas Taleb [100] defines the concept succinctly:

> A derivative is a security whose price ultimately depends on that of another asset (called underlying).

So a financial derivative could, in principle, refer to any legal contract that references something underlying (whether it is traded or not). Peter L. Bernstein [7] writes:

> Derivatives are financial instruments that have no value of their own. That may sound weird, but it is the secret of what they are all about. They are called derivatives because they derive their value from the value of some other asset, which is precisely why they serve so well to hedge the risk of unexpected price fluctuations.

Our job is to figure out how to price these financial contracts, and how to hedge them, failing which they are bets, at best. Bernstein explains [7]:

> Derivatives come in two flavors: as futures (contracts for future delivery at specified prices), and as options that give one side the opportunity to buy from or sell to the other side at a prearranged price.

We shall now consider futures and options.

4.2 FUTURES AND FORWARDS

Forward contracts specify an agreement between two agents on a transaction to deliver an asset, at a specific price (called the delivery price) at an agreed date (typically referred to as the delivery or maturity date). Forwards are generally OTC instruments (so-called OTC-markets); this means the prices could depend on volumes and perceived credit risk, for example.

The pre-specified price, at which the asset (underlying a forward contract) will be delivered, is chosen in such a way that the forward contract has zero value at inception of the contract. The expected fair value of a stock at a certain maturity date, for example, is often referred to as the forward value.

Futures contracts have the same feature but the distinction lies in standardised contracts trading on an exchange – this typically ensures liquidity, and minimised costs, so that futures contracts change hands multiple times during their lifetimes. Importantly, futures are margined; hence, buyers/sellers do not suffer from counterparty credit issues as with forward contracts.

4.3 OPTIONS

An option describes a contractual right (but not the obligation) to buy (or sell) an asset for a fixed price. The following conditions are typically specified:

1. **Strike or Exercise Price:** The contractual purchase or sale price.

2. **Expiry Date:** The date when the derivative contract expires (or matures).

3. **Asset Description:** Examples of the underlying asset could be stocks, indices, currencies, baskets of currencies, commodities, and bonds.

4. **Exercise Features/Styles:** American-style options can be exercised at any time up to the maturity date while European-style options can only be exercised at the maturity date.

The right, but not the obligation, to buy an asset, is referred to as a call option. Because the contract is a right, with no obligation, we can write the payoff of the contract (also known as the intrinsic value of the option) as:

$$(S_T - X)^+ \equiv \max\{0, S_T - X\}$$
$$= \begin{cases} 0, & \text{if } S_T \leq X \\ S_T - X, & \text{if } S_T > X, \end{cases} \quad (4.1)$$

where S_T represents the value of the asset at time T (the option expiration date) and the exercise price is denoted by X. See, Figure 4.1.

In reality, a right, with no obligation, sounds too good to be true. The catch? It will cost you!

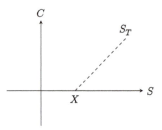

FIGURE 4.1   Payoff: Call option with strike X.

So, the correct payoff for a strategy will have to take account of the initial cost of the call option (the so-called option premium), say c, adjusted for carry, that is,

$$\max\{0, S_T - X\} - ce^{rT}, \tag{4.2}$$

using a continuously compounded interest rate, r, as defined in Chapter 3.

Remark: It is important to understand that we seldom hold options to their expiration. The premium, c, reflects the value of the contingent claim and becomes a tradable asset. This asset yields the opportunity to gain(or lose) moves of the underlying, volatility, and interest rates, through changes in the value of the claim.

Correspondingly, an option that gives you the right, but not the obligation, to sell an asset, is called a put option and the payoff is given by:

$$(X - S_T)^+ \equiv \max\{0, X - S_T\}$$
$$= \begin{cases} 0, & \text{if } S_T \geq X \\ X - S_T, & \text{if } S_T < X. \end{cases} \tag{4.3}$$

Here is some sagely advice from Howard Marks [76]. He provides us with an appreciation of risk, return, and investment opportunities, specifically in the context of options:

> Because the future is inherently uncertain, we usually have to choose between (a) avoiding risk and having little or no return, (b) taking a modest risk and settling for a commensurately modest return, or (c) taking on a high degree of uncertainty in pursuit of substantial gain but accepting the possibility of substantial permanent loss. Everyone would love a shot at earning

big gains with little risk, but the "efficiency" of the market – meaning the fact that the other participants in the market aren't dummies – usually precludes this possibility.

The focus of this book is on the pricing (i.e., determining the value of c) and corresponding hedging of contingent claims. See, Chapters 8 and 10, and, in particular, Sections 8.3 and 10.4. In Chapter 16, we discuss so-called exotic options, where we outline a range of conditions to the contractual choices described above.

Contingent claims are not only used in the financial markets – 4.4, we discuss so-called real options.

4.4 REAL OPTIONS

In many practical applications, we frequently search for embedded optionality in products or projects. The term real options is often used to describe investment situations involving non-financial, that is, real assets together with some degree of optionality. In essence, unlike financial options that involve the right to buy or sell financial assets at predetermined prices, real options involve tangible assets or operational flexibility within a business setting.

Real options provide a framework for assessing the value of flexibility embedded within investment opportunities, allowing decision-makers to adapt their strategies in response to evolving market conditions.

Luenberger [72] provides a succinct discussion on real options:

> In fact, it is possible to view almost any process that allows control as a process with a series of operational options. These operational options are often termed **real options** to emphasize that they involve *real* activities or *real* commodities, as opposed to purely financial commodities, as is the case, for instance, of stock options.

Consider the following examples involving real options:

1. **Expansion Option:** A retail chain that considers expanding into new markets may have the option to scale its operations up or down, based on the performance of initial stores, thereby minimising exposure to volatility in market pricing.

2. **Investment Timing Option:** Consider a technology company considering the launch of a new product. It has the option to defer the launch to capitalise on emerging market trends, or technological advancements, thereby maximising potential returns.

3. **Abandonment Option:** An energy company investing in a new oil exploration project retains the option to abandon the venture if geological surveys indicate unfavourable prospects, thereby limiting potential losses.

    Similarly, the owner of an oilfield has an option to drill for oil that she may exercise at any time. In fact, since she can drill for oil in each time period, she actually holds an entire series of options. On the other hand, if she only holds a lease on the oilfield that expires on a specified date, then she holds only a finite number of drilling options.

4. **Switching Option:** An automotive manufacturer investing in alternative fuel technologies may have the option to switch production between electric and traditional vehicles based on regulatory changes, or shifts in consumer preferences, ensuring adaptability to market dynamics.

5. **Phased Investment Option:** A construction firm undertaking a large infrastructure project may have the option to proceed in phases, assessing market conditions and project feasibility at each stage before committing further resources.

6. **Licensing Option:** A pharmaceutical company developing a new drug may have the option to license the technology to other firms in different geographic regions, diversifying revenue streams and reducing market risk.

7. **Contractual Option:** An outsourcing firm entering into a long-term contract with a client may negotiate contractual clauses that provide the option to renegotiate terms or terminate the agreement under specified conditions, thereby preserving flexibility in a dynamic business environment.

Real options are exploited through strategic decision-making processes aimed at maximising flexibility and minimising downside risks. Strategies for exploiting real options include:

1. **Flexibility in Timing:** Deferring irreversible commitments until uncertainties are resolved or favourable conditions emerge.
2. **Strategic Expansion:** Investing in projects with embedded options to expand, contract, or switch strategies based on changing market dynamics.
3. **Value Enhancement:** Incorporating the value of real options explicitly into investment appraisal techniques, such as decision trees or Monte Carlo simulations.

Valuation of real options involves quantitative models that account for factors such as volatility, time to expiration, and the value of underlying assets. Techniques such as the binomial option pricing model (see, Section 8.3) or the Black–Scholes model (see, Section 12.4) adapted for real assets are commonly used to estimate the value of embedded options.

Real options offer decision-makers a powerful framework for navigating uncertainty and seizing opportunities in dynamic business environments. By recognising and exploiting the strategic flexibility inherent in investment decisions, firms can enhance their competitiveness and drive sustainable growth.

CHAPTER 5

Option Strategies

Options have a long and checkered history. Once again the Bible (Genesis 29) contains the earliest reference to a business option. The incident occurred when Jacob wished to marry Rachel, younger daughter of Laban. Laban agreed, provided that Jacob first pay him with seven years of labor. After that period, Jacob would have an option on Rachel's hand. One can see that options were already off to a bad start because Laban reneged on the contract and delivered to Jacob his elder daughter, Leah, instead.

– BURTON G. MALKIEL [75]

Prices, like everything else, move along the line of least resistance. They will do whatever comes easiest.

– JESSE LIVERMORE (1877–1940)

We always plan too much and always think too little.

– JOSEPH A. SCHUMPETER (1883–1950)

The market is a device for transferring money from the impatient to the patient.

– WARREN BUFFETT

5.1 INTRODUCTION

We introduced the basics of financial options in Section 4.3 recollect: options represent rights, with no obligation, apart from paying some premium to receive the associated right.

We now want to create an understanding of our profit/loss (so-called P/L) profile from owning these rights. This will assist us in understanding the risk/reward profile of different trading strategies.

In essence, we have the following terminal P/L, π_T, for basic positions, that is, positions involving a long/short position in the stock or an option, only:

1. **Long/(Short) Stock:**

$$\pi_T = \pm S_T \mp S_0 e^{rT}.$$

   In words, if we bought (i.e., went long) a stock at price S_0, our profit (or loss) is determined by the current price, S_T, less the original price adjusted for carry cost, that is, $S_0 e^{rT}$.

   The term 'short' refers, generally, to selling an asset that you do not own.

2. **Long/(Short) Call Option:**

$$\pi_T = \pm (S_T - X)^+ \mp c e^{rT}.$$

   Selling (or, more correctly, writing) a call with no hedge is frequently called a naked call write. Correspondingly, selling a call against a long stock holding is called a covered call strategy.

3. **Long/(Short) Put Option:**

$$\pi_T = \pm (X - S_T)^+ \mp p e^{rT}.$$

   Writing a put with no hedge is called a naked put write. Buying a put against a portfolio is called a protective put purchase.

Note, in each equation, the latter part reflects the cost of establishing the position, adjusted for interest carry, that is, $c e^{rT}$ and $p e^{rT}$. This is easy enough, but the next step entails that we need to get a sense of the costs involved. Fred Schwed [99] summarises:

Options are infinitely attractive to dream about. We all know many stocks which have moved much more than ten points in a month, and more than fifty points in three months. But when a man stops dreaming these transactions and tried doing them something different always seems to happen.

The customer who buys some options soon discovers that his costs are considerably higher than at first appears.

In Section 5.2, we provide details of basic Vanilla European option costs.

5.2 VANILLA OPTIONS

We use the Black–Scholes [9] and Merton [79] pricing model for a European call option. The model is derived in Chapters 10 and 12. See, Equation (10.12), for example.

We consider an index, denoted by S_t, with volatility σ (see, Section 5.3). Here, we provide the general option pricing formula where the index pays continuous dividends at rate q. We value the option at time t, its expiry is given by T, with strike price X. The price of a European call option is given by:

$$C(S_t, X, r, q, \sigma, T) = S_t e^{-q(T-t)} N(d_1) - X e^{-r(T-t)} N(d_2), \quad (5.1)$$

where $N(\cdot)$ denotes the cumulative Normal distribution function; see, Equation (17.3). Furthermore,

$$d_1 = \frac{\ln(S_t/X) + (r - q + \sigma^2/2)(T-t)}{\sigma\sqrt{T-t}},$$

and

$$d_2 = d_1 - \sigma\sqrt{T-t}.$$

In 5.1 below, we list prices for a range of European call options by making use of this formula. These are obtained for a fixed spot price, $S_0 = 100$ and for the purpose of illustration using a volatility of $\sigma = 20\%$. Note, we frequently refer to $T - t$ as the tenor of an option.

We have used interest and dividend rates at zero, that is, $r = d = 0\%$. This is done to illustrate the time value of the options only.

TABLE 5.1 European Call Prices: $S_0 = 100$, $\sigma = 20\%$, $r = d = 0\%$

Tenor/Strike	70	80	90	100	102.5	105	107.5	110	120	125
0.25	30.00	20.04	10.71	3.99	2.91	2.06	1.42	0.95	0.15	0.05
0.50	30.02	20.31	11.77	5.64	4.54	3.62	2.85	2.21	0.72	0.39
0.75	30.10	20.72	12.73	6.90	5.81	4.85	4.02	3.31	1.43	0.90
1.00	30.25	21.19	13.59	7.97	6.88	5.91	5.05	4.29	2.15	1.48
2.00	31.16	23.08	16.41	11.25	10.18	9.20	8.29	7.47	4.83	3.85
5.00	34.45	27.82	22.27	17.69	16.69	15.74	14.84	13.99	11.03	9.78

TABLE 5.2 European Call Prices: $S_0 = 100$, $\sigma = 30\%$, $r = d = 0\%$

Tenor/Strike	70	80	90	100	102.5	105	107.5	110	120	125
0.25	30.04	20.40	12.02	5.98	4.88	3.95	3.16	2.50	0.89	0.50
0.50	30.34	21.43	13.99	8.45	7.36	6.39	5.52	4.75	2.50	1.78
0.75	30.84	22.50	15.61	10.34	9.26	8.28	7.38	6.57	4.04	3.13
1.00	31.43	23.53	17.01	11.92	10.86	9.88	8.98	8.14	5.44	4.42
2.00	33.92	27.12	21.44	16.80	15.79	14.83	13.93	13.08	10.13	8.90
5.00	40.26	34.88	30.25	26.27	25.36	24.49	23.66	22.85	19.91	18.60

It is very instructive to note the effect of the volatility parameter, σ, on the prices of options. In Table 5.2, we show the same options as in Table 5.1 but change the volatility from $\sigma = 20\%$ to $\sigma = 30\%$.

5.3 ABOUT VOLATILITY

In the original derivation of the formula, Myron Scholes and I made the following unrealistic assumptions: A stock's volatility is known, and never changes.

– FISCHER BLACK

Taleb [100] provides a succinct definition of volatility:

Volatility is best defined as the amount of variability in the returns of a particular asset.

Let us demonstrate how we would calculate the volatility of an asset. Assume $\{P_i\}$ denotes a price series, such as the daily closing values of an equity index. We estimate the variance by calculating

$$V = \frac{1}{n-1} \sum_{i=1}^{n} (X_i - \bar{X})^2, \tag{5.2}$$

where we use

$$X_i = \ln(P_i/P_{i-1}),$$

and

$$\bar{X} = \frac{1}{n}\sum_{i=1}^{n} X_i.$$

Throughout, n denotes the number of observations in the sampling horizon we are considering. The volatility, σ, is calculated by annualising V as follows:

$$\sigma = \begin{cases} \sqrt{250V}, & \text{for daily data, assuming 250 trading days per year} \\ \sqrt{52V}, & \text{for weekly data} \\ \sqrt{12V}, & \text{for monthly data.} \end{cases} \quad (5.3)$$

We distinguish between several related volatility concepts:

1. **Actual or Realised Volatility:** The realised volatility is a measure of the actual movement experienced by the market. Using Equations (5.2) to (5.3) above, with $n = 23$ would give us the current realised one-month volatility, for example.

2. **Implied Volatility:** The volatility, σ, observed in markets where options do trade, is called the implied volatility, that is, the volatility parameter used in the Black–Scholes–Merton option pricing model, Equation (5.1), that yields the correct market price.

   This description from Derman and Miller [40] is instructive:

   > In finance we refer to backed-out estimates of the future values of parameters obtained by forcing a market price to fit a model as implied values. Implied values are predictions, but they are predictions based on currently observed market prices. The implied volatility can be fruitfully regarded as the market's expected value of future volatility.

3. **Historical Actual Volatility:** We can calculate the realised volatility, for specified time periods, over time, as experienced by the market.

We typically use this as a time series. We could, for example, calculate the time series of 23-day historical volatility over a time period.

A useful analysis is typically to look at historical volatility levels for different sample-periods, say 23-days, 66-days, 125-days, and 250-days, representing historical one-month, three-month, six-month, and one-year historical volatility. We then look at the range of historical volatility levels to give us a sense of the volatility risk of an underlying. This is typically referred to as a volatility cone.

4. **Historical Implied Volatility:** Implied volatility is related to the value of options, and traders will look at the historical time series thereof for trading purposes and technical analysis, for example.

5. **Future Actual Volatility:** We want to hedge contingent claims. Our methodology is to create a continuously re-balanced hedge, using the asset and cash. The strategy exposes us to the realised volatility of the asset over time; hence, the future actual volatility matters to us.

6. **Future Implied Volatility:** We typically trade options on implied volatility; hence, the future implied volatility matters as it creates trading opportunities to buy or sell options as trading assets.

Jan Loeys [70] provides the following fundamental look at volatility, which provides an interesting conclusion to this section:

> Market volatility is not a mystery but should be thought of as fundamental volatility, of growth, earnings, inflation, plus technical forces which are largely due to leverage, positions, market plumbing and such. Another way of looking at vol is as a function of the number of shocks and surprises hitting the system, the propagation and contagion forces around them (mostly leverage) and the shock absorbers that counteract them (largely central banks).

5.4 TERM STRUCTURE OF VOLATILITY

We start by discussing a simplified one-day move model to illustrate some volatility-based concepts. This toy-model is instructive as it provides an example of using implied volatility and variance. It also allows us the opportunity to introduce the term-structure of volatility.

Suppose we have two options trading on an underlying stock.

Notably:

A: Option with n_1 days to maturity and implied volatility σ_1.

B: Option with n_2 days to maturity and implied volatility σ_2.

We write our stock variance as a combination of some baseline volatility, $\bar{\sigma}$, and a one-day specific jump-event, σ_J, which is common to both options, and assume $n_1 < n_2$. Therefore, because variance is additive, we can write

$$n_1 \sigma_1^2 = (n_1 - 1)\bar{\sigma}^2 + \sigma_J^2, \qquad (5.4)$$
$$n_2 \sigma_2^2 = (n_2 - 1)\bar{\sigma}^2 + \sigma_J^2. \qquad (5.5)$$

Subtracting (5.4) from (5.5), we solve for $\bar{\sigma}$,

$$\bar{\sigma}^2 = \frac{n_2 \sigma_2^2 - n_1 \sigma_1^2}{n_2 - n_1},$$

from which σ_J follows as:

$$\sigma_J^2 = \frac{(n_1 - 1)(n_2 - 1)}{(n_2 - n_1)} \left(\frac{n_1}{n_1 - 1} \sigma_1^2 - \frac{n_2}{n_2 - 1} \sigma_2^2 \right).$$

Note the requirement here that

$$\left(\frac{n_1}{n_1 - 1} \right) \sigma_1^2 - \left(\frac{n_2}{n_2 - 1} \right) \sigma_2^2 \geq 0,$$

failing which the model is mis-specified or an arbitrage opportunity is available.

The baseline move is given by

$$\sqrt{\bar{\sigma}^2/252}$$

while the specific jump-event move is given by

$$\sqrt{\sigma_J^2/252}.$$

So we retain two numbers that resemble one-day percentage moves (a bit more accurately, these are log-returns); using the model assumptions (we assume, effectively, the stock's daily returns are each observations of a random variable; the distribution of the random variable is the same for all days except one), these numbers represent the fair expectation of a one-day move and the average daily move over the rest of the period.

Practically, the option market has therefore given us an estimate of the expected move on an earnings-announcement day, for example; as investors, we could create a trading strategy based on our (differing) views.

It is useful to understand the forward variance as well, that is, if we denote the variance between period n_1 and n_2 by $\sigma_{1,2}$, we can write, using the same logic as we had in Equation (5.4), for example,

$$\sigma_{1,2}^2 = \sqrt{\frac{n_2 \sigma_2^2 - n_1 \sigma_1^2}{n_2 - n_1}}, \qquad (5.6)$$

provided that

$$n_2 \sigma_2^2 - n_1 \sigma_1^2 \geq 0.$$

We could therefore construct a volatility function $\sigma(t)$ as

$$\sigma(t) = \begin{cases} \sigma_1, & \text{for } t \leq n_1 \\ \sigma_{1,2}, & \text{for } n_1 < t \leq n_2, \end{cases}$$

which will yield the average volatilities for the two options. We therefore find a so-called term-structure of implied volatilities for the stock.

5.5 STRATEGIES

By observing Tables 5.1 and 5.2, we realise that options are not necessarily cheap. We might, however, only desire to gain upside exposure to a share price movement over a range of prices and consequently cheapen the cost of optionality. This leads us to consider various option strategies. Various names exist for different option trading strategies. We describe a few examples below:

1. **Long Straddle:** This strategy involves the simultaneous purchase of a put and call option with the same strike and maturity. Payoff is given by:

$$\max\{0, S_T - X, X - S_T\} - (p + c)e^{rT},$$

where p and c denote the initial price of the put and call options, respectively. A straddle position clearly expresses a view on volatility of the underlying.

2. **Long Strangle:** This involves the simultaneous purchase of a put (strike X_p) and call (strike X_c), with $X_p < X_c$. Payoff is given by:

$$\max\{0, S_T - X_c, X_p - S_T\} - (p+c)e^{rT}.$$

Strangles are typically used to express a view on volatility of the underlying.

3. **Call Ratio Spread:** In essence, the simultaneous purchase of a call, strike X_1, and sale of n calls with strike $X_2 > X_1$. Payoff then:

$$\max\{0, S_T - X_1\} - n\max\{0, S_T - X_2\} + (nc_2 - c_1)e^{rT}.$$

Ratio spread trades are frequently done to achieve gearing. A strong view can be expressed that the underlying will move up or down with no initial capital outlay – this will clearly be punished (as a function of the ratio used) if the view is incorrect. The strategy therefore expresses a strong view on an outcome within a specified range.

4. **Bull Spread:** The bull spread involves the simultaneous purchase of a call with strike X_l and sale of a call with strike $X_h > X_l$, that is, payoff,

$$\max\{0, S_T - X_l\} - \max\{0, S_T - X_h\} - (c_l - c_h)e^{rT}.$$

So, essentially, a specific case of the ratio call spread. A bear spread could be defined equivalently to take a view on the underlying losing value. Bull and bear spreads could be traded using put and call options.

5. **Butterfly:** The strategy is created by taking long positions in calls with strike X_1 and X_3 and a short position in twice the number of calls with strike X_2. Note that $X_1 < X_2 < X_3$ and that $X_2 = \frac{1}{2}(X_1+X_3)$ (Figure 5.1).

6. **Calendar Spread:** This is a position created by taking a long position in a call option with maturity T_1 and a corresponding short position in a call with maturity T_2 with $T_1 < T_2$.

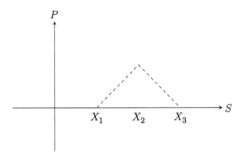

FIGURE 5.1 Payoff: Butterfly strategy, strikes $X_2 - X_1 = X_3 - X_2$.

5.6 PACKAGES

The following structures (or packages) exist and are typically sold to clients:

1. **Zero Cost Collar:** Client purchases a put, strike X_p, and sells a call, strike $X_c > X_p$, struck so that there is no up-front premium payable. The payoff of the options without the underlying investment is given by:

$$\max\{0, X_p - S_T\} - \max\{0, S_T - X_c\}.$$

This package is frequently used by fund managers looking to protect assets at no *initial* cost. (Unwind costs, based off market moves, could be material.) The collar is typically used as a hedge overlay to protect a long share basket. Collars are also referred to as risk reversals.

2. **Collar:** $(X_2 > X_1)$

$$\min\left[\max\{S_T, X_1\}, X_2\right].$$

The collar entails a long stock position with an overlay of a long put at strike X_1 and a short call at strike X_2.

5.7 RISKS AND CONSIDERATIONS

It should be clear from the discussion above that one can achieve various desired payoff profiles by making use of combinations of ordinary put and call options.

Derivatives therefore allow us the ability to profit from a correct view of the distribution of an underlying asset's return at the maturity date.

It is important, however, to take the following practical risks and operational aspects (among others) into account:

1. **Risk Appetite:** Derivatives strategies entail some form of leverage, we need to ensure our understanding thereof through the 'life-cycle' of a strategy.

2. **Scope Creep:** There are many different ways to achieve very seemingly similar outcomes using derivative strategies. It is easy (and often tempting) to change option strategies but we need to ensure we stick to our original mandate if we are not option traders. We need to understand the system requirements as well as the accounting aspects of the strategies we embed into our business.

3. **Settlement:** Option strategies could settle in physical assets or cash; it is important to understand the settlement process and ensure its consistency with a desired strategy.

4. **Liquidity:** We might not always want to hold a derivative strategy to maturity. It is therefore important to understand whether we would be able to find a market-maker to help us exit a specific strategy. Far out-of-the-money options, for example, or customised exotic OTC derivatives could be difficult to trade.

5. **Time Horizon:** Options have expiration dates. The expiration has a considerable influence on your strategy and introduces some execution risk.

6. **Margin:** Option strategies are frequently implemented on exchanges and will therefore require some form of margining. This introduces a level of path-dependency into your strategy and could create significant liquidity constraints on a portfolio which should be taken into account.

7. **Valuation:** Options are typically marked-to-market in any professional context. This means we report and consider the valuation of an option periodically. Because options are geared instruments, the differences between reporting periods could be significant. This means that the gearing effect of derivatives becomes an asset/liability as well.

8. **Basis Risk:** Our option strategy could be based on an index while we are trying to hedge a basket of shares, for example. This will entail some basis risk that could have adverse hedging consequences. In this context, basis risk refers to the mismatch between the index and the basket of shares, a measure of which can be calculated by the so-called tracking-error volatility measure or a Value-at-Risk (VaR) calculation.

5.8 FURTHER READING

Most introductory texts on financial derivatives provide some discussion on option trading strategies. It is important to bear in mind that, in the context of a portfolio of derivative securities, the exposure to a combination of different variables is more important. Consult Rebonato [92] or Taleb [100].

The basic option strategies, however, are described by the following authors, for example, Chance [25], Hull [58], Ritchken [93], and Wilmott [104].

CHAPTER 6

Basic Option Bounds

You know the joke about two economists walking down the street and seeing a $20 bill lying on the sidewalk.

The first economist says, "Look at that $20 bill." The second says, "That can't really be a $20 bill lying there, because if it were, someone would have picked it up already."

So they walk on, leaving the $20 bill undisturbed.

– J. PFEFFER

6.1 INTRODUCTION

In this book, we shall discuss some of the methods used to model, value, and hedge derivatives. Using the wise insights of Emanuel Derman [38]:

> Price is what it costs to buy a security. Value is what you think it is worth. The difference between the two is what modeling and investing are about.
>
> Worth is a mental quality, a matter of opinion and therefore subject to uncertainty. In contrast to the value of the electron's mass or charge, there is nothing absolute about the value of a financial asset.

It is, therefore, important to understand that we consider securities on a relative basis. We start by considering fundamental relationships between financial options (so-called contingent claims) and their underlying spot instruments. These relationships involve bounds derived by using only no-arbitrage assumptions; hence, we make no assumptions regarding the movement of the underlying spot instruments. See the original work by Merton [79], as well as [11, 25, 61, 92].

These option bounds provide valuable insight to understand the relationships between an option and its underlying variables as well as carry cost considerations. They are also useful when testing numerical approximation techniques to value options, for example.

6.2 BASIC ASSUMPTIONS

We shall make the following assumptions for the purposes of the present discussion:

1. **Asset:** Assume that we are working with a common stock as underlying asset. We assume that the stock pays no dividends.

2. **Trading:** Assume trading without any frictions (i.e., no taxes and no transaction costs).

3. **Constraints:** Assume we can hold positions with no liquidity or balance sheet constraints.

4. **Unwind:** We have no forced unwinding of any positions.

5. **Arbitrage:** Assume no-arbitrage opportunities.

We shall work with the *law of one price* throughout this analysis. The *law of one price* states if two securities produce identical payouts in all future states, then to avoid riskless arbitrage, their current prices must be the same. Emanuel Derman explains this fundamental concept quite succinctly [37]:

> Financial economists grandiosely refer to this law as the *law of one price,* which states that securities with identical future payouts, no matter how the future turns out, should have identical prices. It's the essential – perhaps the only – principle of the field.

6.3 ELEMENTARY RELATIONS: CALL OPTIONS

We shall first establish some basic results for European call options using no-arbitrage type arguments. We shall use the notation,

$$C(S_t, T; X),$$

to denote a European call option on an underlying S_t time, t, with strike, X, and maturity, T. Similar notation holds for a European put option, namely $P(S_t, T; X)$.

C0: The value of a call option is always non-negative, that is,

$$C(S_t, T; X) \geq 0, \qquad (6.1)$$

for $S_t \geq 0$.

The proof follows by noting that options represent a right, not an obligation. Therefore, the contract value has to be non-negative. □

Hence, once you have paid for an option, you, as the buyer, have no further liability. Do not confuse that with P/L, or exposure, though.

C1: If the stock price is zero, then the value of a call option must be zero, that is,

$$C(0, T; X) = 0.$$

This is an important special-case. Asset prices typically have zero as a limiting case – a company at zero value, typically, has ceased to exist.

C2: A call option can never be worth more than the underlying reference asset, that is,

$$C(S_t, T; X) \leq S_t. \qquad (6.2)$$

If the option were trading at a higher level than the underlying spot, one would sell the option and buy the underlying, ensuring a profit when the option matures and, hence, an arbitrage opportunity. □

C3: The minimum value of a European call option is given by

$$C(S_t, T; X) \geq \max\{0, S_t - B(t, T)X\}, \qquad (6.3)$$

where $B(t,T)$ is the price of a ZCB, maturing at time, T, that has a payout of one for certain.

We can prove this relation by considering we hold a call, $C(S_t, T; X)$, and X discount bonds, $B(t,T)$. At the maturity date

$$C(S_T, T; X) + XB(T,T) = \begin{cases} S_T, & \text{if } S_T \geq X, \\ X, & \text{if } S_T < X, \end{cases}$$

which is always larger than S_T. By using the *law of one price*,

$$C(S_t, T; X) + XB(t,T) \geq S_t,$$

which yields our result by recalling Equation (6.1). □

1. Following from the results of Equations (6.2) and (6.3), we can therefore also write:

$$\max\{0, S_t - B(t,T)X\} \leq C(S_t, T; X) \leq S_t, \quad (6.4)$$

We note that these bounds are sharp; that is, they can be attained in practice.

2. An interesting, direct consequence of Equation (6.4) is that the value of a perpetual call, $C(S, T; X)$, with $T \to \infty$, has to be $C = S$ (provided that rates are non-negative).

3. Another useful consequence of Equation (6.4) is that:

$$0 \leq \partial_S C \leq 1.$$

C4: If $X_1 < X_2$, then

$$C(S_t, T; X_1) \geq C(S_t, T; X_2). \quad (6.5)$$

We can prove this by considering $C(S_t, T; X_1) - C(S_t, T; X_2)$. At the option maturity date,

$$C(S_T, T; X_1) - C(S_T, T; X_2) = \begin{cases} 0, & \text{if } S_T \leq X_1 \\ S_T - X_1, & \text{if } X_1 < S_T < X_2 \\ X_2 - X_1, & \text{if } S_T \geq X_2 \end{cases}$$

$$\geq 0.$$

Our result follows from the *law of one price*. □

This means that call option prices are a decreasing function of the strike X. In essence,

$$\partial_x C(S_t, T; X) \leq 0,$$

which follows directly from Equation (6.5). Therefore, by making use of Equation (6.3), we can write:

$$-B(t, T) \leq \partial_x C(S_t, T; X) \leq 0. \tag{6.6}$$

These bounds are sharp and attained in the limits as $X \to 0$ and $X \to \infty$ whenever $t < T$.

C5: The difference between the prices of otherwise identical European calls cannot exceed the present value of the difference between their strike prices, that is, given $X_1 < X_2$:

$$C(S_t, T; X_1) - C(S_t, T; X_2) \leq (X_2 - X_1) B(t, T). \tag{6.7}$$

We can prove this by considering $C(S_t, T; X_1) - C(S_t, T; X_2) - (X_2 - X_1) B(t, T)$. At the maturity date,

$$C(S_T, T; X_1) - C(S_T, T; X_2) - (X_2 - X_1)$$
$$= \begin{cases} -(X_2 - X_1), & \text{if } S_T < X_1 \\ S_T - X_2, & \text{if } X_1 \leq S_T < X_2 \\ 0, & \text{if } S_T \geq X_2 \end{cases}$$
$$\leq 0.$$

Our result follows from the *law of one price*. □

C6: By making use of Equations (6.5) and (6.7), we can write:

$$C(S_t, T; X_2) \leq C(S_t, T; X_1) \leq C(S_t, T; X_2) + B(t, T)(X_2 - X_1).$$

Realising that $B(t, T)$ does not depend on X_1 or X_2, this result is equivalent to the mathematical statement that $C(S_t, T; X)$ is a Lipschitz continuous function of X, with Lipschitz constant $B(t, T)$.

C7: Given that $t \leq T_1 \leq T_2$,

$$C(S_t, T_1; X) \leq C(S_t, T_2; X).$$

This result is a consequence of Equation (6.3) and application of the *law of one price* as stated above. □

C8: Call option prices are homogeneous in the spot/strike price, that is,

$$C(\lambda S, \lambda X, T) = \lambda C(S, X, T). \qquad (6.8)$$

The result can be seen by expanding the payoff at time T as

$$\max\{0, \lambda(S_T - X)\} = \lambda \max\{0, S_T - X\}.$$

The result follows from the *law of one price* as before. □

An interesting consequence of Equation (6.8) can be seen by differentiating the expression implicitly with respect to λ. By setting $\lambda = 1$, we obtain the following Euler relation:

$$S \partial_S C + X \partial_X C = C. \qquad (6.9)$$

Continuing, we can also show that

$$S \partial_S^2 C = -X \partial_{SX}^2 C$$
$$S \partial_{XS}^2 C = -X \partial_X^2 C.$$

Consequently,

$$\partial_S^2 C = \left(\frac{X}{S}\right)^2 \partial_X^2 C. \qquad (6.10)$$

We shall make further use of these relations in Chapter 12.

Remark: We need to understand what the effects are of incorporating taxes (depending on the jurisdiction and distinctions between so-called capital and revenue accounts), transaction costs, as well as balance sheet costs (typically these entail regulatory restrictions, or capital requirements, or liquidity dependence) in our analyses. In many cases, at the very least, the profitability (or viability) of establishing an arbitrage transaction is in question.

6.4 EUROPEAN PUT–CALL PARITY

We shall now consider the relationship between European put and call options using arbitrage type arguments. For European options, the following relationship holds between put and call options (with similar strikes):

$$P(S_t, T; X) = C(S_t, T; X) + XB(t, T) - S_t. \qquad (6.11)$$

Suppose we form a portfolio consisting of a European call option, a short position in the stock, S_t, and a holding in X discount bonds, $B(t, T)$. Therefore, at the time of expiry, that is, $t = T$, we have:

$$C(S_T, T; X) + XB(T, T) - S_T = \begin{cases} 0, & \text{if } S_T \geq X \\ X - S_T, & \text{if } S_T < X \end{cases}$$
$$= P(S_T, T; X).$$

Equation (6.11) follows by using the *law of one price* as before. □

This extremely important relationship is referred to as put–call parity for European options. Don Chance [25] cites this entertaining example of the early use of put–call parity:

> In the mid 1800s, famed New York financier Russel Sage began creating synthetic loans using the principle of put–call parity. Sage would buy the stock and a put from his customer and sell the customer a call. By fixing the put, call, and strike prices, Sage was creating a synthetic loan with an interest rate significantly higher than usury laws allowed.

Here are a couple of uses, among others, of put–call parity:

1. **Pricing Consistency:** Examining Equation (6.11), we see that we are effectively 'tying' a relationship between the spot, cash, and options markets.

2. **Synthetic Positions:** Put–call parity allows synthetic creation of positions. For example, being long a put and short a call, at the same strike, yields a short position in the spot market.

3. **Testing:** Suppose we have written a numerical approximation to price options. Using put–call parity we can test for fundamental violations or perform relative accuracy tests.

4. **Rates Impacts:** Put–call parity demonstrates a direct link between pricing and rates.

5. **Option Strategies:** Put–call parity assists us in designing, and understanding, option strategies. It also helps us establish relationships between the sensitivities of options to underlying factors. (See, Chapter 15.)

To illustrate the last point, let us differentiate Equation (6.11) with respect to S to obtain

$$\partial_S P = \partial_S C - 1.$$

This very useful relationship shows us the exposure of a put is the same as holding a call and being short of the underlying. However, differentiating with respect to S again, we have that

$$\partial_S^2 P = \partial_S^2 C.$$

We shall provide a full discussion of option sensitivity measures, or so-called 'Greeks', in Chapter 15.

Remarks:

1. **Voting Rights:** We need to be careful – buying a call and selling a put, all with the same strike, provides a synthetic long exposure; however, we do not retain any voting rights of the underlying shares, for example. Hence, certain physical rights are not necessarily conferred or implied.

2. **Taxes:** It is also important to understand, in practice, that put–call parity could be violated depending on the specific taxation regime.

6.5 ELEMENTARY RELATIONS: PUT OPTIONS

We shall now establish some basic results for European put options using arbitrage type arguments. We shall also use the put–call parity results.

P0: The value of a put option is always non-negative, that is,

$$P(S_t, T; X) \geq 0.$$

Again – once you've paid for an option, you, as the buyer, have no further liability. Don't confuse that with P/L or exposure, though.

Put options are frequently used for so-called event-insurance purposes. It is instructive to note that we can create put options using call options and short stock positions, however.

P1: If the stock price is zero, then the value of a put option must be its discounted exercise price, that is,

$$P(S_t = 0, T; X) = B(t, T)X.$$

P2: A put option can never be worth more than the strike price, that is,

$$P(S_t, T; X) \leq X.$$

Correspondingly,

$$P(S_t, T; X) \leq B(t, T)X. \tag{6.12}$$

P3: The minimum value of a European put option is given by

$$P(S_t, T; X) \geq \max\{0, B(t, T)X - S_t\}. \tag{6.13}$$

Therefore, from Equations (6.12) and (6.13), we have that:

$$\max\{0, B(t, T)X - S_t\} \leq P(S_t, T; X) \leq B(t, T)X.$$

P4: If $X_1 < X_2$, then

$$P(S_t, T; X_1) \leq P(S_t, T; X_2).$$

Using put–call parity:

$$\begin{aligned}
&P(S_t, T; X_1) - P(S_t, T; X_2) \\
&= (C(S_t, T; X_1) - S_t + X_1 B(t, T)) - (C(S_t, T; X_2) \\
&\quad - S_t + X_2 B(t, T)) \\
&= (C(S_t, T; X_1) - C(S_t, T; X_2)) - (X_2 - X_1) B(t, T) \\
&\leq 0.
\end{aligned}$$

The last line follows directly from the relationship in Equation (6.7). □

P5: The difference between the prices of otherwise identical European puts cannot exceed the present value of the difference between their strike prices, that is, given $X_1 < X_2$:

$$P(S_t, T; X_2) - P(S_t, T; X_1) \leq (X_2 - X_1)B(t, T).$$

We use put–call parity again. Consider:

$$\begin{aligned} &P(S_t, T; X_2) - P(S_t, T; X_1) \\ &= (C(S_t, T; X_2) - S_t + X_2 B(t, T)) - (C(S_t, T; X_1) \\ &\quad - S_t + X_1 B(t, T)) \\ &= (C(S_t, T; X_2) - C(S_t, T; X_1)) - (X_2 - X_1)B(t, T) \\ &\leq (X_2 - X_1)B(t, T). \end{aligned}$$

The last observation follows directly from Equation (6.5). □

P6: Suppose $X_1 < X_2$. Then,

$$P(X_1, T) < \frac{X_1}{X_2} P(X_2, T). \tag{6.14}$$

To prove the result, we consider two portfolios – the aim is to evaluate their outcomes at the expiry date and invoke the law of one price. Portfolio A consists of put option $X_2 P(X_1)$ whereas portfolio B consists of $X_1 P(X_2)$ options. Consider the Arbitrage Table 6.1 (for time T, i.e., at expiry).

Now note that:

$$X_2(X_1 - S_T) - X_1(X_2 - S_T) = S_T(X_1 - X_2) < 0.$$

This means that portfolio A will always be worth less than portfolio B; the result follows from application of the *law of one price* as before. □

P7: Put option prices are homogeneous in the spot/strike price, that is,

$$P(\lambda S, \lambda X, T) = \lambda P(S, X, T).$$

TABLE 6.1 Arbitrage Table for Equation (6.14)

Portfolio	$S_T < X_1$	$X_1 \leq S_T < X_2$	$S_T \geq X_2$
Portfolio A	$X_2(X_1 - S_T)$	0	0
Portfolio B	$X_1(X_2 - S_T)$	$X_1(X_2 - S_T)$	0

CHAPTER 7

Relations Between Options

The derivatives we call options, by expanding the variety of risks that can be insured, help to create Kenneth Arrow's ideal world where all risks are insurable.

– PETER L. BERNSTEIN (1919–2009)

I believe you can summarize the essence of quantitative finance on one leg, too: "If you want to know the value of a security, use the price of another security that's as similar to it as possible. All the rest is modeling. Go and build."

– EMANUEL DERMAN [37]

If finance is about anything, it is about the messy world we inhabit.
It's best to learn axioms only after you've acquired intuition.

– EMANUEL DERMAN

To understand derivatives models it is essential to grasp how they are put to use. In general, a pricing model can be of interest to plain-vanilla-option traders, to relative-value traders and to complex-derivatives traders. Relative-value and plain-vanilla traders are interested in models because of their ability to predict how option prices should move relative to the underlying, and relative to each other, given a certain *move in the underlying. For both these classes of user, models should therefore have not just a descriptive, but also a prescriptive dimension.*

— RICCARDO REBONATO [92]

7.1 INTRODUCTION

The aim of this chapter is to continue with the results of Chapter 6. We shall now consider the relationship between European put and call options with different strike prices using no-arbitrage type arguments. We also show a very interesting consequence of the relationships and bounds we have derived, which is fundamental to our understanding of pricing methods for derivative securities.

7.2 STRIKE PRICE RELATIONS: PUT AND CALL OPTIONS

Let $X_1 \leq X_2 \leq X_3$ be three strike prices. Consider three European call option prices, namely $C(X_1), C(X_2)$, and $C(X_3)$. The call options are identical in all respects except for their strike prices. For simplicity, we start by considering (see, Figure 5.1 as well)

$$X_3 - X_2 = X_2 - X_1.$$

Then, to prevent riskless arbitrage to be established, we must have that

$$C(X_2) \leq (C(X_1) + C(X_3))/2. \qquad (7.1)$$

This relationship is proven below for a more general case; we note it holds equally well for put options. Note that Equation (7.1) implies that

$$\partial_x^2 C \geq 0.$$

The relationship given above could also be generalised. Specifically, option prices are convex in the exercise price. That is, given $X_1 < X_2 < X_3$, we could write

$$C(X_2) \leq \lambda C(X_1) + (1-\lambda)C(X_3), \qquad (7.2)$$

where

$$\lambda = (X_3 - X_2)/(X_3 - X_1). \qquad (7.3)$$

Let us emphasise this result – options are a convex function of the exercise price X.

TABLE 7.1 Arbitrage Table for Equation (7.2)

Portfolio	$S_T < X_1$	$X_1 \leq S_T < X_2$	$X_2 \leq S_T < X_3$	$S_T \geq X_3$
Portfolio A	0	0	$S_T - X_2$	$S_T - X_2$
Portfolio B	0	$\lambda(S_T - X_1)$	$\lambda(S_T - X_1)$	$\lambda(S_T - X_1) + (1-\lambda)(S_T - X_3)$

To prove the result, we consider two portfolios. Our aim is to evaluate their outcomes at the expiry date and invoke the law of one price. Portfolio A consists of call option $C(X_2)$ whereas portfolio B consists of $\lambda C(X_1) + (1-\lambda)C(X_3)$ options. Consider Arbitrage Table 7.1 (for time T, i.e., at expiry):

Note that:

$$(S_T - X_2) - \lambda(S_T - X_1)$$
$$= \{(S_T - X_2)(X_3 - X_1) - (X_3 - X_2)(S_T - X_1)\}/(X_3 - X_1).$$
$$= \{(S_T - X_3)(X_2 - X_1)\}/(X_3 - X_1) < 0,$$

whenever $X_2 \leq S_T < X_3$. Similarly,

$$(S_T - X_2) - \{\lambda(S_T - X_1) + (1-\lambda)(S_T - X_3)\} = 0.$$

This proves the relationship in Equation (7.2) as it becomes evident that the value of portfolio A is always less than or equal to the value of portfolio B at the expiry date. The proof is completed by virtue of the *law of one price* as used previously. □

7.3 AN APPLICATION OF BUTTERFLY SPREADS

Let us consider a butterfly spread for a fixed maturity T:

$$BF(X, \alpha) = \frac{C(X - \alpha) - 2C(X) + C(X + \alpha)}{\alpha^2}.$$

Let us consider the limit as $\alpha \to 0$:

$$\lim_{\alpha \to 0} BF(X, \alpha) = \partial_X^2 C(X) \geq 0.$$

If we define $p(X, T)$ as follows,

$$p(X, T) := e^{rT} \partial_X^2 C(X), \tag{7.4}$$

we find that

$$\int_0^\infty p(X,T)dX = 1. \tag{7.5}$$

This can be seen by considering

$$\int_0^\infty p(X,T)dX = e^{rT}\left[\partial_X C(X)\right]_0^\infty,$$

and using Equations (6.3) and (6.6), which yield

$$\lim_{X\to 0} \partial_X C = -e^{-rT},$$

and

$$\lim_{X\to\infty} \partial_X C = 0,$$

hence producing Equation (7.5). □

We can use this result to value options! Consider:

$$\partial_X \int_0^\infty (S-X)^+ p(S,T)dS = -\int_X^\infty p(S,T)dS.$$

This follows from Leibniz' rule for differentiation of an integral. Therefore,

$$-\partial_X \int_X^\infty p(S,T)dS = p(X,T),$$

that is,

$$e^{-rT}\partial_X^2 \int_0^\infty (S-X)^+ p(S,T)dS = e^{-rT}p(X,T) = \partial_X^2 C(X).$$

This means that

$$C(X,T) = e^{-rT}\int_0^\infty (S-X)^+ p(S,T)dS. \tag{7.6}$$

Technically, the integral expression and the call price could differ by a linear function of X. Such a linear function would be equal to zero. This can be seen by noting that:

$$\int_0^\infty Xp(X,T)dX = Se^{rT}, \qquad (7.7)$$

which follows from integration by parts. Noting the relations (6.2) and (6.3), therefore, as well as the fact that

$$\lim_{X\to\infty} C(S_T, X) = 0,$$

means any such linear functional needs to be zero. □

Equation (7.6) will be used repeatedly in derivatives pricing. Refer in particular to Section 12.2 where this result, with Equation (7.7), proves to be of vital importance to derive the governing Black–Scholes PDE used to determine the prices of contingent claims or derivatives.

Equation (7.7), specifically, shows that the distribution defined by Equation (7.5) is a risk-neutral distribution. Risk neutrality implies that the expected total return on the asset equals the risk-free rate [16].

7.4 REPLICATING GENERAL EUROPEAN PAYOFFS

Let us use $p(X, T)$ defined in Equation (7.4) to calculate the value of an option with an arbitrary payoff, namely $f(S_T)$, at the maturity date T. We can suppress the T dependence for ease of notation. Equation (7.6) holds more generally; following Breeden and Litzenberger [12], by making use of Equation (7.4), we write the discounted expectation as

$$\begin{aligned} f(S_0) &= e^{-rT} E[f(S_T)] \\ &= e^{-rT} \int_0^\infty f(X) p(X) dX \qquad (7.8) \\ &= \int_0^\infty f(X) \partial_X^2 C(X) dX. \end{aligned}$$

Assuming f is twice continuously differentiable, using integration by parts, we can rewrite Equation (7.8) as

$$\begin{aligned} f(S_0) &= -f(0)\partial_X C(0) - \int_0^\infty \partial_X f \partial_X C dX \\ &= -f(0)\partial_X C(0) + \partial_X f(0) C(0) + \int_0^\infty C \partial_X^2 f dX. \end{aligned} \qquad (7.9)$$

We used the fact that European call options with infinite strikes are worthless to establish Equation (7.9). Furthermore, we have that

$$C(0) = S,$$

and also that

$$\partial_X C(0) = -e^{-rT},$$

following the reasoning from before.

Hence, we can complete our analysis and write Equation (7.9) as follows:

$$f(S_0) = f(0)e^{-rT} + \partial_X f(0)S_0 + \int_0^\infty C \partial_X^2 f dX. \qquad (7.10)$$

So, our option with payoff $f(S_T)$ can be hedged, using $f(0)$ ZCBs $\partial_X f(0)$ zero-strike calls, that is, the spot, S_0, and the continuum of call options in quantities $\partial_X^2 f(X) dX$. Practically, only a finite number of strikes would be available, hence the result would be approximated.

Remark: We can obtain Equation (7.10) by other arguments as well. Using the Dirac delta function, $\delta(\cdot)$, we could write

$$f(S) = \int_0^\infty f(X)\delta(S - X) dX.$$

Using integration by parts yields

$$f(S) = f(0) + f'(0)S + \int_0^\infty f''(X)(S - X)^+ dX.$$

Hence, Equation (7.10) follows by virtue of us assuming no arbitrage and using the *law of one price*. (Note we have abused notation by suppressing the time dependence.)

Further details are contained in, for example, Albanese and Campolieti [3], Carr and Madan [22], Carr and Wu [23], and Gatheral [50]. This model-free result is frequently used to derive the pricing of volatility derivatives; for example, see the discussions in Derman and Miller [40] and Gatheral [50].

CHAPTER **8**

Binomial Pricing Model: I

There are only 10 types of people in the world: those who understand binary, and those who don't.

– ANONYMOUS

On a binomial tree, prices move like knights on a chessboard, one discrete step forward in time and up or down a notch in price. Binomial trees are easy to draw and, in a jerky way, mimic the behavior of real prices or indices. As the grid of the chessboard becomes progressively finer, prices move more and more continuously – they start to diffuse, in fact – and the binomial model becomes equivalent to the Black-Scholes model.

– E. DERMAN [37]

[Upon proving that the best betting strategy for "Gambler's Ruin" was to bet all on the first trial.] It is true that a man who does this is a fool. I have only proved that a man who does anything else is an even bigger fool.

– JULIAN LOWELL COOLIDGE (1873–1954)

All models are wrong; some are useful.

– GEORGE E.P. BOX (1919–2013)

8.1 INTRODUCTION

Markets go up and markets go down. Simple, isn't it?!

In the binomial setup, the stock, $S(t)$, follows the simple process,

$$S(t + \delta t) = \begin{cases} uS(t), & \text{with probability } p, \\ dS(t), & \text{with probability } (1-p), \end{cases} \quad (8.1)$$

where, for now, we consider u and d to be specified positive numbers. The probability p denotes a subjective probability of the specified stock's moves and determines the expected value of the stock, that is,

$$E^p[S(t + \delta t)] = puS(t) + (1-p)dS(t). \quad (8.2)$$

We also refer to this as a *subjective* or *real-world* expectation.

If we think about two potential scenarios of a stock price being given, with subjective probabilities, it will be the stock-price expected value that we differ on between different parties.

The following view from Jan Loeys [70] provides us with insight on thinking about asset prices and their changes:

> The starting point of Finance is the Theorem of **Market Efficiency** which posits that under ideal conditions what we all know should be in the price. Only new information moves the price. Hence, it is changes in expectations about the future that drive asset prices, not the level of anything.

He also provides this insight [70]:

> There is a fundamental difference between an asset price and a forecast. A forecast is a single outcome that you consider the most likely, among many. In statistics, we call this the mode. An asset price, in contrast, is closer to the probability-weighted mean of the different scenarios you consider possible in the future.

8.2 HOW DO WE PREVENT ARBITRAGE?

In essence, arbitrage opportunities are presented when we create a portfolio consisting of derivative securities and cash, which have a non-negative payoff in all future states of the market, with a non-positive initial outlay. Let us examine this statement.

If we consider the binomial setup described above, the stock, $S(t)$, follows the process,

$$S(t + \delta t) = \begin{cases} uS(t), & \text{with probability } p, \\ dS(t), & \text{with probability } (1-p), \end{cases}$$

and we can therefore form a portfolio, $V(t)$, which consists of $\alpha S(t)$ and β in the money market account (earning a continuously compounded interest rate, r). Therefore, at time, $t + \delta t$, we have:

$$\begin{aligned} V(t + \delta t) &= \alpha S(t + \delta t) + \beta e^{r\delta t} \\ &= \begin{cases} \alpha\, uS(t) + \beta e^{r\delta t}, & \text{with probability } p, \\ \alpha\, dS(t) + \beta e^{r\delta t}, & \text{with probability } (1-p). \end{cases} \end{aligned}$$

Note, that we have assumed that fractional stock holdings are in order.

Suppose we want to test for arbitrage. At time, t, we need to have $V(t) = 0$, which means that we require

$$\beta = -\alpha S(t),$$

that is,

$$V(t + \delta t) = \begin{cases} \alpha S(t)\left[u - e^{r\delta t}\right], & \text{with probability } p, \\ \alpha S(t)\left[d - e^{r\delta t}\right], & \text{with probability } (1-p). \end{cases}$$

Arbitrage over a single time-step would mean that

$$P(V(t + \delta t) > 0) = 1,$$

hence, to avoid arbitrage, we need to require that:

$$(u - e^{r\delta t}) \gtreqless 0,$$

while

$$(d - e^{r\delta t}) \lesseqgtr 0.$$

Consequently, to avoid arbitrage, we require that

$$d < e^{r\delta t} < u. \tag{8.3}$$

A consequence of Equation (8.3) is that $e^{r\delta t}$ is a linear combination of u and d. Therefore, owing to the absence of arbitrage, we can write

$$e^{r\delta t} = qu + (1-q)d,$$

for a value, q, which we will interpret as the so-called risk-neutral probability of a move up. Hence, solving for q, we find

$$q = \frac{e^{r\delta t} - d}{u - d}. \tag{8.4}$$

Note, using the probability, q, we have:

$$\begin{aligned}
e^{-r\delta t} E^q \left[S(t + \delta t) \right] &\equiv e^{-r\delta t} \left[qu\, S(t) + (1-q)d\, S(t) \right] \\
&= e^{-r\delta t} \left[\frac{e^{r\delta t} - d}{u - d} uS(t) + \frac{u - e^{r\delta t}}{u - d} dS(t) \right] \\
&= e^{-r\delta t} S(t) \left[\frac{(u - d)e^{r\delta t} - du + ud}{u - d} \right] \\
&= S(t).
\end{aligned} \tag{8.5}$$

Hence, using the risk-neutral probability, q, the discounted expectation of the future asset value is equal to today's asset value.

8.3 BINOMIAL OPTION PRICING

Suppose we want to replicate the value of a call option. If we were to hedge a written call option using a full allocation to the underlying, we would be fully hedged if the stock moves up, but have problems if the stock moves down. Similarly, if we hold no hedge, we would be very happy if the stock price falls, but lose money if it increases.

We could, therefore, ask whether we could find some 'optimal' hedge value between 0 and 1. Practically, this would mean we hold a portion, Δ, in the stock, S, and an amount, B, in the money market account, to wit:

$$C = \Delta S + B. \tag{8.6}$$

We denote $C_u = C(uS(t), t+\delta t)$ and $C_d = C(dS(t), t+\delta t)$. Therefore, we require

$$C_u = \Delta uS + Be^{r\delta t} \qquad (8.7)$$
$$C_d = \Delta dS + Be^{r\delta t}. \qquad (8.8)$$

We, therefore, have two equations, with two unknowns, which lead to an easy solution. Subtracting (8.8) from (8.7) yields a solution for Δ, namely

$$\Delta = \frac{C_u - C_d}{S(u-d)},$$

and B follows by substituting Δ into either of Equations (8.7) or (8.8), namely

$$B = e^{-r\delta t} \frac{uC_d - dC_u}{u - d}.$$

Contingent claims can, therefore, be hedged in this market that is complete. In terms of Equation (8.6), we therefore have

$$\begin{aligned} C &= \Delta S + B \\ &= \frac{C_u - C_d}{u - d} + e^{-r\delta t} \frac{uC_d - dC_u}{u - d} \\ &= e^{-r\delta t} \left[\frac{e^{r\delta t} - d}{u - d} C_u + \frac{u - e^{r\delta t}}{u - d} C_d \right]. \end{aligned} \qquad (8.9)$$

We note the presence of q as defined in Equation (8.4). Also note the absolute absence of p. We can therefore rewrite the price of the option as:

$$C = e^{-r\delta t}[qC_u + (1-q)C_d]. \qquad (8.10)$$

Equivalently, we could write

$$\begin{aligned} C(S(t), t) &= e^{-r\delta t}\left[q\, C(uS(t)) + (1-q)\, C(dS(t))\right] \\ &= e^{-r\delta t} E^q[C(S(t+\delta t), t+\delta t)]. \end{aligned} \qquad (8.11)$$

In other words, the value of a contingent claim can be expressed as the discounted expected value of the payoff evaluated using the risk-neutral measure (as implied by the risk-neutral probability).

We recall the work to establish Equations (8.10) and (8.11). It is important to understand that using the risk-neutral measure carries the implication that the contingent claim will be hedged in a specific manner.

8.4 DISCUSSION OF ASSUMPTIONS

Mark Rubinstein, of Cox–Ross–Rubinstein (CRR) renown [34], wrote:

> At one point, we wondered how it was that the then two-year-old Black-Scholes approach to valuing options could recreate a riskless payoff using only the option and its underlying asset. It was then that Sharpe said, "I wonder if it's really that there are only two states of the world, but three securities, so that any one of the securities can be replicated by the other two".

Recall the basic assumptions we made in Section 6.2. Additionally, we now add some important assumptions.

1. **Probability:** Our model of the stock assumed moves with specified probabilities, whereas these subjective probabilities play no role in the hedging, or pricing, of the contingent claim. The hedging programme we adopted, therefore, creates the basis for parsimonious pricing and hedging.

2. **Trade:** We are assuming throughout that we can trade the underlying (long or short) on a fractional basis.

3. **Money Market:** We have access to a money market account, free of default.

4. **Funding:** We have unlimited borrowing or lending at the same interest rate. In Section 11.7, we discuss an extension of the binomial pricing model to allow collateralised funding.

Chriss [28] provides the following insight on binomial tree methods:

> What binomial models make apparent about stock price models in general is that all future possibilities of the stock price are spelled out ahead of time. These models really are models for future price movement. Every possibility, those that happen, as well as those that do not, is mapped out ahead of time, and

moreover, we can say exactly what the probability of each stock price occurring is, once we know all the transition probabilities.

He concludes:

Naturally, no one actually expects stocks to behave as in a binomial model. Rather, what is hoped is that the model captures some of the essential probabilistic features of the underlying stock, at least enough for options pricing.

Note that we provide a general discussion of pertinent option pricing assumptions in Section 11.3. In Section 11.7, specifically, we show an extension of the binomial method to funding curves.

8.5 CHOICE OF PARAMETERS

If we return to Equation (8.1), we recall that we used u and d as given. In this section, we provide one derivation (but note, many derivations are possible) for u and d. From our previous analysis, we recall Equation (8.4)

$$quS + (1-q)dS = Se^{r\delta t}.$$

Hence,

$$q = \frac{e^{r\delta t} - d}{u - d}.$$

On our tree, we also want to assess the variance of the underlying asset. We use

$$\text{Var}[Q] = E[Q^2] - E[Q]^2.$$

Therefore,

$$qu^2 + (1-q)d^2 - (qu + (1-q)d)^2 = \sigma^2 \delta t, \qquad (8.12)$$

which we rewrite as

$$q = \frac{e^{2r\delta t} + \sigma^2 \delta t - d^2}{u^2 - d^2}. \qquad (8.13)$$

We make use of a so-called recombining tree; mathematically that means

$$u = \frac{1}{d}. \tag{8.14}$$

Combining Equations (8.4), (8.13), and (8.14), we find upon simplification that

$$1/d + d = e^{-r\delta t}[1 + e^{2r\delta t} + \sigma^2 \delta t]$$
$$\equiv 2A.$$

Therefore,

$$d^2 - 2Ad + 1 = 0,$$

and using Taylor's theorem to simplify A, we find upon approximation that

$$A \approx 1 + \sigma^2 \delta t/2 + \ldots$$

Solving the quadratic equation, we have

$$d = A \pm \sqrt{A^2 - 1}$$
$$\approx 1 \pm \sigma\sqrt{\delta t} + \sigma^2 \delta t/2 + \ldots \tag{8.15}$$
$$\approx e^{\pm \sigma\sqrt{\delta t}}.$$

Remark: We approximated the variance of the binomial stock process by $S^2\sigma^2\delta t$ in Equation (8.12). This approximation is further justified using the results in Section 10.2, Equation (10.6).

8.6 RELATIVE ASSET PRICES

In Equation (8.5), we showed that

$$e^{-r\delta t} E^q [S(t + \delta t)] = S(t),$$

when using the probability q given by Equation (8.4), that is,

$$q = \frac{e^{r\delta t} - d}{u - d}.$$

We note that we can rewrite Equation (8.5) as follows:

$$E^q\left[\frac{S(t+\delta t)}{e^{r\delta t}}\right] = S(t).$$

Now, recalling Equation (3.2), we abuse notation and note that we can write the money market account dynamics as

$$M(t+\delta t) = M(t)e^{r\delta t},$$

for constant rates. Therefore,

$$E^q\left[\frac{S(t+\delta t)}{M(t+\delta t)}\right] = \frac{S(t)}{M(t)}. \tag{8.16}$$

A process, whose expected future value is equal to its current value, is called a *martingale*, see, [19, 83]. For example, the relative prices are martingales under the measure q in this example.

It is now a fun question to ask whether we can find a measure, given a probability, $\bar{q}$, such that

$$E^{\bar{q}}\left[\frac{M(t+\delta t)}{S(t+\delta t)}\right] = \frac{M(t)}{S(t)}.$$

This would require solving for $\bar{q}$ in

$$\bar{q}\frac{e^{r\delta t}}{uS} + (1-\bar{q})\frac{e^{r\delta t}}{dS} = \frac{1}{S},$$

which implies that

$$\bar{q} = \frac{u - ude^{-r\delta t}}{(u-d)}. \tag{8.17}$$

Using $\bar{q}$ to price a call option, for example, we require

$$E^{\bar{q}}\left[\frac{C}{S}\right] = \bar{q}\frac{C_u}{uS} + (1-\bar{q})\frac{C_d}{dS},$$

so that

$$\begin{aligned}
C &= \bar{q}\frac{C_u}{u} + (1-\bar{q})\frac{C_d}{d} \\
&= \left[\frac{u(1-de^{-r\delta t})}{u(u-d)}\right]C_u + \left[\frac{d(ue^{-r\delta t}-1)}{d(u-d)}\right]C_d \quad (8.18) \\
&= e^{-r\delta t}\left[\frac{e^{r\delta t}-d}{u-d}C_u + \frac{u-e^{r\delta t}}{u-d}C_d\right].
\end{aligned}$$

Hence, we recover the option pricing formula, namely Equation (8.10), from before, and we recall that it was derived using the measure implied by the probability, q, as shown in Equation (8.4).

8.7 LIMITING BEHAVIOUR

Limits are the silent architects of mathematical continuity, guiding us through the infinitesimal to grasp the immense.

– ISAAC NEWTON (1643–1727)

In this section, we investigate the limiting behaviour of the binomial method as the discretisation parameter, δt, becomes arbitrarily small.

Consider an arbitrary derivative $f(S,t)$. Recall the backwards recursion we use in the binomial method, namely

$$f(S,t) = e^{-r\delta t}[qf(uS, t+\delta t) + (1-q)f(dS, t+\delta t)], \quad (8.19)$$

as given by Equations (8.10) and (8.11) in Section 8.3.

To aid our investigation, we appeal to Taylor's theorem applied to $f(uS, t+\delta t)$ and $f(dS, t+\delta t)$, that is,

$$f(uS, t+\delta t) = f(S,t) + (u-1)S\partial_s f + \frac{1}{2}(u-1)^2 S^2 \partial_s^2 f + \partial_t f \delta t, \quad (8.20)$$

and

$$f(dS, t+\delta t) = f(S,t) + (d-1)S\partial_s f + \frac{1}{2}(d-1)^2 S^2 \partial_s^2 f + \partial_t f \delta t. \quad (8.21)$$

Substituting Equations (8.20) and (8.21) into the pricing relationship Equation (8.19) yields

$$e^{r\delta t} f(S,t) - f(S,t) = \delta t \, \partial_t f + [q(u-1) + (1-q)(d-1)] \, S \partial_s f$$
$$+ \frac{1}{2} \left[q(u-1)^2 + (1-q)(d-1)^2 \right] S^2 \partial_s^2 f. \tag{8.22}$$

By making use of Equation (8.4), we can write

$$q(u-1) + (1-q)(d-1) = e^{r\delta t} - 1$$
$$= r\delta t + O(\delta t^2), \tag{8.23}$$

and we used Taylor's theorem to write the last line. Furthermore, using Equation (8.12), we have

$$q(u-1)^2 + (1-q)(d-1)^2 = \sigma^2 \delta t + (e^{r\delta t} - 1)^2$$
$$= \sigma^2 \delta t + O(\delta t^2), \tag{8.24}$$

where the last step again follows by using Taylor's theorem. Therefore, using Equations (8.23) and (8.24), by ignoring terms $O(\delta t^2)$ and cancelling δt terms, we can write Equation (8.22) as

$$rf = \partial_t f + rS \partial_s f + \frac{1}{2} \sigma^2 S^2 \partial_s^2 f. \tag{8.25}$$

This is the celebrated Black–Scholes partial differential equation (PDE). We shall cover the PDE in more detail starting in Chapter 12. See, the original work by Black and Scholes [9] and also Merton [79].

We have, therefore, demonstrated that the binomial option pricing method is consistent with the Black–Scholes PDE, in the limit, as the discretisation parameter, δt, tends to zero.

8.8 FURTHER READING

The binomial method serves as a quick introduction to pricing methods for derivative securities and serves to build good intuition. The original article by CRR [34] is a great start. The work by Chance [26] provides an excellent overview. See, also, [6, 19, 28, 54, 58, 72, 92, 93, 105] as well as the seminal work on extensions to the volatility smile [39, 43] as well as the insights offered by Derman and Miller [40].

CHAPTER 9

Binomial Pricing Model: II

The surest way to remain a winner is to win once, and then not play any more.

– ASHLEIGH BRILLIANT

The fundamental law of investing is the uncertainty of the future.

– PETER L. BERNSTEIN

..., I shall take Leave to lay down this Self-evident Truth: That any one Chance or Expectation to win any thing is worth just such a Sum, as wou'd procure in the same Chance and Expectation at a fair Lay. As for Example, if any one shou'd put 3 Shillings in one Hand, without letting me know which, and 7 in the other, and give me Choice of either of them; I say, it is the same thing as if he shou'd give me 5 Shillings; because with 5 Shillings I can, at a fair Lay, procure the same even Chance or Expectation to win 3 or 7 Shillings.

– CHRISTIAN HUYGENS [36]

The replication of an option payoff by trading in the underlying is the theoretical bedrock of unique pricing by no-arbitrage. Exact option-payoff replication is not necessary in order to price options without *allowing arbitrage. It is however necessary if we want to associate a unique price to an option simply by invoking absence of arbitrage.*

– R. REBONATO [92]

9.1 INTRODUCTION

In this chapter, we shall consider some applications of the binomial option pricing method. We examine some curiosities of the model and also some extensions.

We provide some examples, which show that we obtain arbitrage opportunities in the binomial world if we trade at different prices than implied by the model. We also discuss some alternatives to the CRR [34] parameterisation. We show that some specification of the binomial parameters could lead to the violation of put–call parity.

9.2 BINOMIAL ARBITRAGE EXAMPLE

Let us consider a one-period binomial tree example. We want to value a European call option with the following parameters:

$$
\begin{aligned}
S(0) &= 100 \\
X &= 100 \\
T - t &= 1 \\
r &= 10\% \\
\sigma &= 20\%.
\end{aligned}
$$

We use the values for u and d, derived in Equation (8.15), with $\delta t = 1$, that is,

$$u = e^{\sigma \sqrt{\delta t}} = e^{.2} = 1.2214,$$

and

$$d = 1/u = 0.8187.$$

In Figure 9.1, we show the stock evolution on the tree as well as the option values, by using the binomial recursion, which was derived in Equation (8.11).

Remember that we derived the binomial option pricing method by showing that we can replicate the value of a contingent claim using a position in the underlying and cash. The following example details the mechanics. Note, in both cases examined below, the Delta of the replicated position is

$$\Delta = \frac{22.14}{(122.14 - 81.87)} = 0.5498.$$

It is instructive to consider what happens if an investor wants to trade the option at values different from the value calculated in Figure 9.1. We consider two cases.

First, assume we have someone willing to sell the option at 13.00 while we know the replication cost is 14.25. As shown in Table 9.1, we, therefore, buy the option at a cheaper price than it should be trading, and replicate it using a short position in the stock as well as a money market investment. This leads to an arbitrage profit of $1.38 = 1.25e^{.1}$, that is, the difference between the bought price and fair price, adjusted for interest.

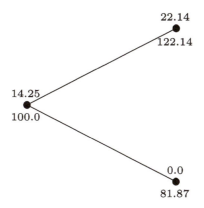

FIGURE 9.1 A two step binomial tree with stock prices (below) and option prices (above) given at each node.

TABLE 9.1 One Period Binomial Arbitrage Example (Buy Option)

Buy call	−13.00	Spot	122.14	P/L option	9.14
Sell delta	54.98			P/L delta	−12.17
Lend	41.98			Interest	4.42
				Profit	1.38
		Spot	81.87	P/L option	−13.00
				P/L delta	9.97
				Interest	4.42
				Profit	1.38

TABLE 9.2 One Period Binomial Arbitrage Example (Sell Option)

Sell call	15.00	Spot	122.14	P/L option	−7.14
Buy delta	−54.98			P/L delta	12.17
Borrow	−39.98			Interest	−4.21
				Profit	0.82
		Spot	81.87	P/L option	15.00
				P/L delta	−9.97
				Interest	−4.21
				Profit	0.82

Second, we assume we have a buyer of the option for 15.00; again, we know the replication value is 14.25. As shown in Table 9.2, we sell the option at a more expensive price than it should be trading and we replicate it using borrowings and physical stock, which again produces an arbitrage profit.

9.3 AN EXAMPLE OF PUT–CALL PARITY VIOLATION

As mentioned in Section 8.5, there are a number of ways to choose the binomial parameters. An interesting (and intuitively appealing) choice of u, d, and probability, q, is given by

$$u = \exp\left((r - \sigma^2/2)\delta t + \sigma\sqrt{\delta t}\right)$$
$$d = \exp\left((r - \sigma^2/2)\delta t - \sigma\sqrt{\delta t}\right)$$
$$q = \frac{1}{2},$$

as described in [29], for example.

Let us consider using this model in a one-period setting to price a synthetic long position, that is, long a call and short a put. Using the binomial recursion, Equation (8.11), we have

$$C(S(t), X) - P(S(t), X) = \exp(-r\delta t)\left[\frac{1}{2}(u+d)S(0) - X\right].$$

By virtue of a Taylor series expansion, we know that,

$$\exp(-r\delta t)(u + d) \approx 2 + \sigma^2 \delta t + \dots$$

Hence, European put–call parity, refer to Equation (6.11), only holds in the limit as $\delta t \to 0$ for this version of the binomial method.

9.4 ALTERNATIVE PARAMETER DERIVATION

Chriss [28] derives an alternative set of values for the binomial method. He considers Equation (8.4) as well as the per-period variance of log-returns:

$$q(1-q)\left(\ln(u/d)\right)^2 = \sigma^2 \delta t.$$

Additionally, he takes

$$q = \frac{1}{2}.$$

The equations we need to solve for u and d, therefore, become

$$u + d = 2\exp(r\delta t)$$
$$u = d\exp(2\sigma \delta t).$$

The solution for u and d follows from direct algebraic manipulation, that is,

$$d = \frac{2e^{r\delta t + 2\sigma\sqrt{\delta t}}}{1 + e^{2\sigma\sqrt{\delta t}}}$$

$$u = \frac{2e^{r\delta t}}{1 + e^{2\sigma\sqrt{\delta t}}}.$$

In terms of the analysis in Section 9.3, we can show that the choice of u and d above does not give rise to violations of put–call parity, for example.

9.5 BINOMIAL METHOD EXAMPLES

Our aim in this section is to consider the binomial option pricing method to price some options using multiple time-steps. We will show the tree (or lattice) and price a European and an American put option.

9.5.1 Methodology

Suppose we wish to value an option over the time interval $[0, T]$. Hence, with a choice of M time-steps, we will have the step-size, $\delta t = T/M$; that is, we denote $t_m = m\,\delta t$, where we will have $m = 0, \ldots, M$ dates. Furthermore, at the m-th time-step of size, δt, we will have $(m+1)$ nodes labelled by an index $n = 0, \ldots, m$.

Following the notation in [105], in a general, two-dimensional binomial tree, or lattice, we will then have the $(m+1)$ potential stock prices, at the m-th time-step, given by

$$S_n^m = S_0 u^n d^{m-n}, n = 0, \ldots, m \qquad (9.1)$$

where $S_0^0 = S_0$, represents the time of option valuation, and u and d are typically given by the CRR parameters in Equation (8.15). Note, however, that various other methods are also possible.

Continuing the style of notation in Equation (9.1), we denote

$$f_n^m = V(S_n^m, t_m), \qquad (9.2)$$

where $t_m = m\,\delta t$ as defined above. We can now use this notation to write the binomial recurrence in Equation (8.15), namely

$$f_n^m = e^{-r\delta t}\left(q f_{n+1}^{m+1} + (1-q) f_n^{m+1}\right), \qquad (9.3)$$

noting that the current value of a derivative will be given by f_0^0. The final time condition, that is, $m = M$, is given by the payoff, namely

$$f_n^m = (X - S_n^M)^+, \qquad (9.4)$$

for a European put option, for example.

Equations (9.1), (9.3), and (9.4) fully describe the binomial method. Practically, we can think of the method with the tree-building process as stepping forward in time. The option price is then calculated by backwards-recursion, using Equation (9.3), after calculating the intrinsic value, Equation (9.4).

Consider, for example, the three-period recombining binomial tree in Figure 9.2. We have simplified the nodes by making use of the recombining feature $ud = 1$.

9.5.2 European Put Option

We now move to an implementation example and consider parameters:

$$\begin{aligned}
S(0) &= 100 \\
X &= 100 \\
T - t &= 1 \\
M &= 3 \\
\delta t &= 1/3 \\
r &= 10\% \\
\sigma &= 20\% \\
u &= 1.1224 \\
d &= 0.8909.
\end{aligned} \qquad (9.5)$$

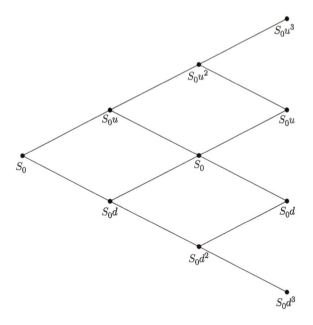

FIGURE 9.2 A three-step recombining binomial tree with $ud = 1$ and stock prices given at each node.

Note, the values for u and d are calculated using the CRR, [34], methodology derived in Equation (8.15), that is,

$$u = e^{\sigma\sqrt{\delta t}} = e^{0.2\sqrt{1/3}} = 1.1224,$$

and

$$d = 1/u = 0.8909.$$

In Figure 9.3, subject to the parameters u and d given in Equation (9.5), we show the evolution of the stock price (Figure 9.4).

9.5.3 American Put Option

We can fairly easily accommodate American options in the binomial recurrence, namely Equation (9.3). The methodology will remain consistent; however, an adjustment needs to be made to account for the possibility of an early exercise. The valuation methodology is now given by

$$f_n^m = \max\left[(X - S_n^m)^+, e^{-r\delta t}\left(q f_{n+1}^{m+1} + (1-q) f_n^{m+1}\right)\right], \quad (9.6)$$

for an American put option, as an example.

94 ■ A Concise Introduction to Financial Derivatives

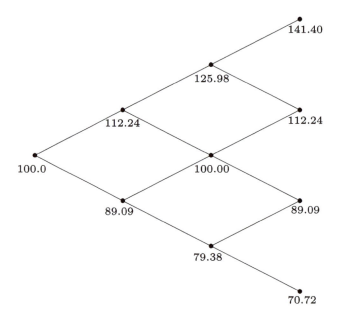

FIGURE 9.3   A three-step binomial tree with stock prices given at each node subject to parameters in Equation (9.5).

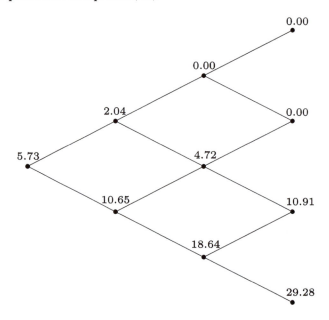

FIGURE 9.4   A three-step binomial implementation of a European put option subject to parameters in Equation (9.5).

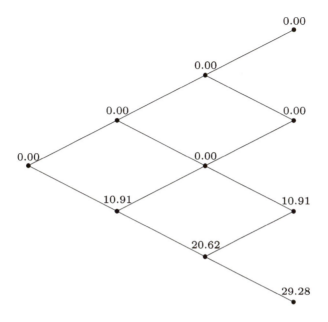

FIGURE 9.5 A binomial tree with three time-steps showing intrinsic value of a put payoff, subject to parameters in Equation (9.5).

In Figure 9.5, we show the intrinsic value of the American put option and we show the calculated value of the American put option using the adjusted formula, namely Equation (9.6), in Figure 9.6.

9.6 BINOMIAL CONVERGENCE

Our aim in this section is to show the convergence characteristics of the binomial method. Very simply, we show the error between the binomial method and the analytical option price. In Figure 9.7, we show the convergence for an at-the-money European call option over various time-steps. In Figure 9.8, we show the same calculations for an option that is 10% out-of-the-money.

9.7 FURTHER READING

In addition to the references provided in Chapter 8, here are some papers with practical details. The work by Chance [26] provides an excellent practical overview. See, also, [2, 28, 29, 66, 85, 104]. The discussion in Wilmott *et al.* [105] is concise.

96 ■ A Concise Introduction to Financial Derivatives

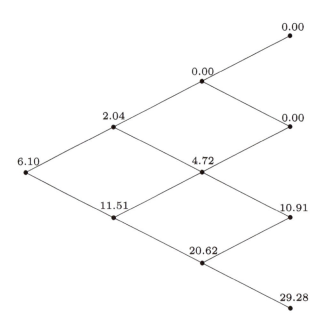

FIGURE 9.6 A three-step binomial implementation of an American put option, subject to the parameters in Equation (9.5).

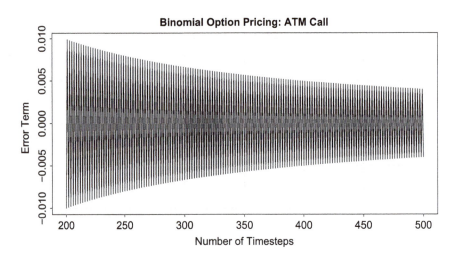

FIGURE 9.7 Error in binomial method as a function of the number of time-steps: At-the-money European call.

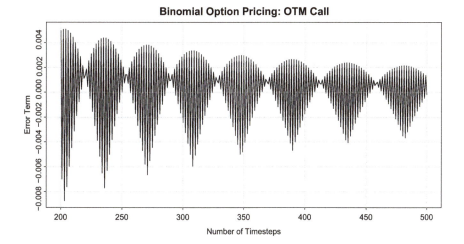

FIGURE 9.8   Error in binomial method as a function of the number of time-steps: Out-of-the-money European call.

Higham [54] wrote a beautiful pedagogic paper on different ways to implement the binomial method, complete with Matlab code. Furthermore, Higham [55] demonstrates this fundamental insight:

> ..., we see that the binomial method corresponds to using an explicit finite difference method on a transformed version of the Black–Scholes PDE.

CHAPTER 10

Option Values: I

A formula can be very simple, and create a universe of bottomless complexity.

– BENOIT MANDELBROT (1924–2010)

Option contracts have been traded for centuries. It is salutary to realize how sophisticated the financial markets were long ago. Richard Dale's book The First Crash describes the London market of the late seventeenth and early eighteenth centuries, where forward contracts, put options (called 'puts') and call options (called 'refusals') were actively traded in Exchange Alley. It seems that the British were at the time a nation of inveterate gamblers. One could bet on all kinds of things: for example, one could buy annuities on the lives of third parties such as the Prince of Wales or the Pretender.

– ALISON ETHERIDGE [4]

Despite the criticisms of the formula, traders have refused alternatives because they have learned its limitations. No experienced trader would willingly trade Black-Scholes-Merton for another pricing tool.

– NASSIM NICHOLAS TALEB [100]

On Wall Street you're puting your money where your mouth is, and you have to walk a delicate line between theory and practicality. Simple is beautiful.

– EMANUEL DERMAN

10.1 INTRODUCTION

In Chapters 8 and 9, we showed the intuition behind the binomial method to value derivative securities.

We have, however, also showed that the value of a contingent claim can be expressed as the discounted expected value of the payoff evaluated using the risk-neutral measure (i.e., probability). In this regard, recall Equations (7.6) and (8.11).

We shall now demonstrate how to use these equations to value European options.

10.2 LOG-NORMAL PRICE DYNAMICS

Let us ponder the price dynamics of the underlying asset first. Generally, if we think about the stock market, we observe day-to-day 'noise' as well as some longer-term trend. We are, therefore, motivated to write a difference-type equation for the evolution of a stock price, which is denoted as $S(t)$, as follows:

$$S(t + \delta t) = S(t) + \mu(t, S(t))\,\delta t + \sigma(t, S(t))\, z(t),$$

where

$$z(t) \sim N(0, 1).$$

This means that $z(t)$ is normally distributed with mean 0 and variance 1. We avoid complications at this stage and keep the functions μ and σ as linear functions of S, that is,

$$S(t + \delta t) = S(t) + \mu S(t)\,\delta t + \sigma S(t)\sqrt{\delta t}\, z(t). \qquad (10.1)$$

Dividing by $S(t)$, and taking the natural log, we find

$$\ln S(t + \delta t)/S(t) = \ln\left(1 + \mu \delta t + \sigma\sqrt{\delta t}\, z(t)\right),$$

which, assuming δt is small, becomes

$$\ln S(t + \delta t)/S(t) \approx \left(\mu \delta t + \sigma\sqrt{\delta t}\, z\right) - \frac{1}{2}\left(\mu \delta t + \sigma\sqrt{\delta t}\, z\right)^2 + \ldots, \qquad (10.2)$$

by virtue of the Maclaurin series expansion

$$\ln(1+x) \approx x - x^2/2! + \ldots$$

Therefore, by expanding Equation (10.2), and ignoring terms with a higher order than $O(\delta t)$, we obtain this very useful relationship:

$$\ln S(t+\delta t)/S(t) \sim N\left((\mu - \frac{1}{2}\sigma^2)\delta t, \sigma^2 \delta t\right),$$

which means that we can write

$$S(t+\delta t) = S(t)e^{(\mu-\sigma^2/2)\delta t + \sigma\sqrt{\delta t}\, z(t)}. \tag{10.3}$$

It should also come as no surprise that we can write, in general, that

$$S(T) = S(t)e^{(\mu-\sigma^2/2)(T-t) + \sigma\sqrt{T-t}\, z}. \tag{10.4}$$

This assertion follows from repeated application of Equation (10.3) and the agreeable assumption that we have independent $z(t) \sim N(0,1)$ corresponding to subsequent timesteps involving δt. See the insightful discussion by Higham [55]. We shall now use this relationship to derive the Black–Scholes value of a European call option.

10.3 EXPECTATIONS

Let us first consider evaluating the expected value of $S(T)^M$, for positive integer values of M, based on Equation (10.4). We find that

$$E\left[S(T)^M\right] = \frac{1}{\sqrt{2\pi}} \int_{-\infty}^{\infty} S(0)^M e^{M(\mu-\sigma^2/2)T + M\sigma\sqrt{T}z} e^{-z^2/2} dz$$

$$= \frac{S(0)^M e^{M(\mu-\sigma^2/2)T}}{\sqrt{2\pi}} e^{\frac{1}{2}M^2\sigma^2 T} \int_{-\infty}^{\infty} e^{-(z-M\sigma\sqrt{T})^2/2} dz,$$

which follows from noting that

$$M\sigma\sqrt{T}z - z^2/2 = -\frac{1}{2}\left(z - M\sigma\sqrt{T}\right)^2 + \frac{1}{2}M^2\sigma^2 T.$$

Therefore,

$$E\left[S(T)^M\right] = S(0)^M e^{(M\mu + M(M-1)\sigma^2/2)T}.$$

If we take $M = 1$, for example, we have:

$$E[S(T)] = S(0)e^{\mu T}. \tag{10.5}$$

Furthermore, the variance of $S(T)$ is given by

$$\begin{aligned}\operatorname{Var}[S(T)] &= E[S(T)^2] - E[S(T)]^2 \\ &= S(0)^2 e^{2\mu T}\left(e^{\sigma^2 T} - 1\right).\end{aligned} \tag{10.6}$$

Suppose that $T = \delta t \ll 1$. By virtue of a Maclaurin series expansion of e^x, we find that we can approximate

$$\operatorname{Var}(S(\delta t)) \approx S^2(0)\sigma^2 \delta t.$$

Refer to the discussion in Section 8.5 pertaining to the variance of a binomial tree.

10.4 EUROPEAN CALL VALUE

We showed that the value of a contingent claim can be expressed as the discounted expected value of the payoff evaluated using the risk-neutral measure (i.e., probability). Recall the analysis for Equations (7.6) and (8.11). We shall now use these equations, with the stock model given by Equation (10.4), to value a European call option.

We start by noting the expected value of the stock price in Equation (10.5). By virtue of Equation (8.5), we will require that

$$E[S(T)] = S(0)e^{rT}, \tag{10.7}$$

where r denotes the continuously compounded risk-free rate as before. This will be the case if we consider the altered, 'risk-neutral' stock price process

$$S(T) = S(t)e^{(r-\sigma^2/2)(T-t)+\sigma\sqrt{T-t}\,z}. \tag{10.8}$$

Let us first use this equation to evaluate a rather easy derivative, namely a contract that delivers $S(T)$ at time T. By virtue of Equation (10.5), which is adjusted using $\mu = r$, the price of the derivative, using $t = 0$, will be

$$\begin{aligned}C(S(0)) &= e^{-rT}E[S(T)] \\ &= S(0),\end{aligned}$$

which is exactly what we should be expecting. □

We now turn to evaluate the price of a European call option using Equations (7.6) and (10.8). Recalling the option payoff, as described in Equation (4.1), we will, therefore, want to evaluate the integral:

$$C(S(0), X) = \frac{e^{-rT}}{\sqrt{2\pi}} \int_{-\infty}^{\infty} \left(S(0)e^{(r-\sigma^2/2)T+\sigma\sqrt{T}z} - X\right)^+ e^{-z^2/2} dz, \tag{10.9}$$

again, at time $t = 0$. We note that the integrand is zero whenever $S(T) \leq X$. This simplifies our work a bit. We, therefore, seek an understanding of when would we have $S(T) > X$. We solve for z as follows:

$$S(0)e^{(r-\sigma^2/2)T+\sigma\sqrt{T}z} > X,$$

that is,

$$z > \frac{\ln(X/S(0)) - (r - \sigma^2/2)T}{\sigma\sqrt{T}}.$$

To conform with the normal literature on the valuation of derivatives, we define

$$d_2 := \frac{\ln(S(0)/X) + (r - \sigma^2/2)T}{\sigma\sqrt{T}}. \tag{10.10}$$

Our option value now reads as follows:

$$C(S(0), X) = \frac{e^{-rT}}{\sqrt{2\pi}} \int_{-d_2}^{\infty} \left(S(0)e^{(r-\sigma^2/2)T+\sigma\sqrt{T}z} - X\right) e^{-z^2/2} dz.$$

The integral can be split into two parts – the last part being the easiest to evaluate, notably

$$\frac{e^{-rT}}{\sqrt{2\pi}} \int_{-d_2}^{\infty} X e^{-z^2/2} dz = \frac{X e^{-rT}}{\sqrt{2\pi}} \int_{-\infty}^{d_2} e^{-z^2/2} dz$$
$$\equiv X e^{-rT} N(d_2).$$

Evaluating the first part of the integral is a bit more tricky, but can be simplified by the familiar 'completing the square' trick, that is, we note that

$$\sigma\sqrt{T}z - z^2/2 = -\frac{1}{2}\left(z^2 - 2\sigma\sqrt{T}z + \sigma^2 T\right) + \frac{\sigma^2 T}{2}.$$

The first part of the integral now reads

$$\frac{e^{-rT}S(0)e^{(r-\sigma^2/2)T}}{\sqrt{2\pi}} \int_{-d_2}^{\infty} e^{\sigma\sqrt{T}z} e^{-z^2/2} dz = \frac{S(0)}{\sqrt{2\pi}} \int_{-d_2}^{\infty} e^{-(z-\sigma\sqrt{T})^2/2} dz$$
$$\equiv S(0)N(d_2 + \sigma\sqrt{T}).$$

Therefore, with reference to Equation (10.10), we define

$$d_1 := d_2 + \sigma\sqrt{T}. \qquad (10.11)$$

We can now conclude our pricing exercise and write the celebrated Black–Scholes formula for the value of a European call as:

$$C(S(0), X) = S(0)N(d_1) - Xe^{-rT}N(d_2), \qquad (10.12)$$

with d_1 and d_2 as defined above. □

Evaluating a European put option follows along the same lines of reasoning. As an alternative derivation, using the European call values, we can make use of put–call parity for European options, that is, Equation (6.11), as well as the fact that

$$N(x) = 1 - N(-x),$$

to derive the European put option formula.

10.5 FORWARD STARTING OPTION

In this section, we consider the pricing of a so-called forward starting option. See Chapter 16. The buyer of the option receives the option only at some predetermined time, T_1, in the future. The strike of the option will be set at the ruling spot price, $S(T_1)$, at that time, yielding the following payoff at the expiry T_2:

$$\max\{0, S(T_2) - S(T_1)\}, \qquad (10.13)$$

for $T_1 < T_2$. Hence, using the analysis in Section 10.4, we can evaluate

$$E^Q\left[(S(T_2) - S(T_1))^+\right],$$

at T_1, given $S(T_1)$, as follows:

$$E^Q\left[(S(T_2) - S(T_1))^+\right] = S(T_1)e^{r(T_2-T_1)}N\left(d_2 + \sigma\sqrt{T_2 - T_1}\right) - S(T_1)N(d_2),$$

where we define

$$d_2 := \frac{(r - \frac{1}{2}\sigma^2)}{\sigma}\sqrt{T_2 - T_1}.$$

Now, given $S(t)$, with $t < T_1 < T_2$, we note that

$$E^Q\left[S(T_1)\right] = S(t)e^{r(T_1-t)}.$$

Hence, using iterated expectations, see Björk [19], for example, and discounting the payoff between T_2 and t, we find the value of the forward starting call as:

$$e^{-r(T_2-t)}\left[S(t)e^{r(T_2-t)}N\left(d_2 + \sigma\sqrt{T_2 - T_1}\right) - S(t)e^{r(T_1-t)}N(d_2)\right]$$
$$= S(t)\left[N(d_1) - e^{-r(T_2-T_1)}N(d_2)\right], \qquad (10.14)$$

where

$$d_1 := d_2 + \sigma\sqrt{T_2 - T_1}.$$

10.6 OPTIONS ON FORWARDS/FUTURES

Our analysis can be extended to consider forward and futures contracts. Consider a forward contract,

$$F = Se^{rT}. \qquad (10.15)$$

If we rewrite Equations (10.11) and (10.10), and using (10.15), we have

$$h^{\pm} = \left[\ln\left(\frac{S(0)e^{rT}}{X}\right) \pm \frac{\sigma^2}{2}T\right] / (\sigma\sqrt{T})$$
$$= \left[\ln\left(\frac{F(0)}{X}\right) \pm \frac{\sigma^2}{2}T\right] / (\sigma\sqrt{T}). \qquad (10.16)$$

Furthermore, if we rewrite Equation (10.12), we have

$$\begin{aligned} C &= e^{-rT}\left(F(0)N(h^+) - XN(h^-)\right) \\ &= B(0,T)\left(F(0)N(h^+) - XN(h^-)\right), \end{aligned} \quad (10.17)$$

where $B(0,T)$ is a ZCB, or discount factor, as we defined in Chapter 3. This is the infamous Black-76 [8] value for a call on a forward. We remark the following:

1. **PDE:** Equation (10.17) is a solution to the PDE that we derive in Section 12.4, Equation (12.10). The derivation follows directly from the Black–Scholes PDE, Equation (8.25), that we derive in Section 8.7.

2. **Calibration:** The Black-76 formula, Equation (10.17), reflects far more than just a change in notation. It uses the discount factor, $B(0,T)$, as well as the forward price; both reflect direct markets in which we can trade. In many markets, we trade futures rather than spot instruments.

3. **Uses:** The Black-76 formula is widely used for FX, commodity, interest rate options, as well as options on dividend paying equities.

10.7 APPROXIMATE CALL VALUES

Suppose we strike a call option at-the-money forward, that is, $X = F(0)$. This implies that we can write Equation (10.16) as

$$h^\pm = \left[\pm\frac{\sigma}{2}\sqrt{T}\right],$$

and, by using the Taylor series approximation,

$$N(x) \approx \frac{1}{2} + \frac{x}{\sqrt{2\pi}} - \frac{x^3}{6\sqrt{2\pi}} + \ldots$$

which holds for $x \ll 1$; Equation (10.17) becomes

$$\begin{aligned} C &= B(0,T)F(0)\left(N(h^+) - N(h^-)\right) \\ &\approx B(0,T)F(0)\frac{\sigma\sqrt{T}}{\sqrt{2\pi}} \\ &\approx 0.4 B(0,T)F(0)\,\sigma\sqrt{T}. \end{aligned}$$

Using Equation (10.15), we can, therefore, also write

$$C(S(0), X, 0, T) \approx 0.4 S(0)\, \sigma \sqrt{T},$$

when the strike is at-the-money-forward, which we can use for comparative purposes with the option values provided in Tables 5.1 and 5.2.

10.8 MONTE CARLO SIMULATION

> *"Can you do addition?" the White Queen asked. "What's one and one and one and one and one and one and one and one and one and one?"*
> *"I don't know," said Alice. "I lost count."*
>
> – THROUGH THE LOOKING GLASS. LEWIS CARROLL

As we have seen, the price of a European derivative is the expectation of the derivative's payoff under the risk-neutral expectation. Appealing to the law of large numbers, an average of the payoff values, calculated from repeated draws from the risk-neutral terminal distribution, converges to the derivatives' price. We detail the calculations below.

Recalling Equations (7.6) and (8.11), we can state, by virtue of risk-neutral valuation, that the value of $V(S(t), t)$ is given by:

$$V(S(t), t) = e^{-r(T-t)} E^Q \left[V(S(T), T) \right], \qquad (10.18)$$

where E^Q denotes the expected value evaluated using the risk-neutral probability as before. Realistically, analytical values are not always available for this integral. Therefore, we seek some approximation. This forms the basis of the so-called Monte Carlo simulation method. Consider,

$$\begin{aligned} V(S(t), t) &= e^{-r(T-t)} E^Q [V(S(T), T)] \\ &\approx e^{-r(T-t)} \frac{1}{n} \sum_{i=1}^{n} V\left(S^{(i)}(T), T\right), \end{aligned} \qquad (10.19)$$

where

$$S^{(i)}(T) = S(t) e^{(r - \sigma^2/2)(T-t) + \sigma \sqrt{T-t}\, Z_i}. \qquad (10.20)$$

$\{Z_i\}_{i=1}^{n}$ denotes independent samples from the standard normal distribution, that is, $Z_i \sim N(0, 1)$.

Joshi [63] provides the following sanguine remarks on the Monte Carlo method:

> The worth of this technique is not so much to value ordinary European options, but to price exotic options where alternative methods are not obviously applicable. We would then have to simulate the entire path, approximating it in little time-steps and then computing the final value of the option along the path. Note, however, that if the option involves some choice on the part of the holder, there is still an issue in deciding what decision the holder would make – for example, in the case of an American option, when should the holder exercise.

So, practically, here is the recipe for performing Monte Carlo simulation:

1. **Stock Process:** We need a specification of the stock process. Recall the earlier discussion in Section 10.3. A typical process could be given by Equation (10.1), for example, or Equations (10.3) or (10.4).
   For the specific purposes of valuing derivative securities, we need to look at the so-called risk-neutral price process; thus, we would typically consider the process given by Equation (10.8).

2. **Generate Random Samples:** If we are valuing a European call option, for example, we will need to evaluate Equation (10.8) for the maturity date T. We require $z \sim N(0, 1)$ in Equation (10.8). One way to generate normal variates, see [46, 91], is to use two independent unit rectangular variates, R_1, R_2. Then
$$\sqrt{-2\ln R_1}\sin(2\pi R_2),$$
and
$$\sqrt{-2\ln R_1}\cos(2\pi R_2),$$
provide two independent standard normal variates. This is known as the Box–Muller method.

3. **Path Dependence:** Evaluating an exotic path-dependent derivative will require evaluation of Equation (10.8) at different sub-intervals of $[0, T]$. An average rate option might require p intervals of equal length $\delta t = T/p$.

4. **Average:** Evaluate the payoff of the derivative at T, for example. Then evaluate the mean as in Equation (10.19).

5. **Estimate Accuracy:** It is easy to calculate the standard error of Monte Carlo simulations. See below.

The Monte Carlo method was first applied to derivative security valuation by Phelim Boyle [15] in 1977. In essence, the method finds appeal if we wish to evaluate the integral

$$\int g(y)f(y)dy = \bar{g},$$

over a definite integration range, where $f(y)$ is a probability density function so that

$$\int f(y)dy = 1,$$

over the same integration range. We can obtain an estimate for $\bar{g}$ by picking n random sample values, y_i from the probability density function $f(y)$, that is,

$$\hat{g} = \frac{1}{n}\sum_{i=1}^{n} g(y_i).$$

The standard deviation of the estimate is given by

$$\hat{s}^2 = \frac{1}{n-1}\sum_{i=1}^{n}(g(y_i) - \hat{g})^2.$$

Hence, Monte Carlo simulation can be seen as a quadrature technique, the standard error of which reduces as

$$\frac{1}{\sqrt{n}},$$

which yields a highly robust method, but at an exceedingly slow rate.

Consider the following numerical example. We want to value a European call option, with parameters given by:

$$S(0) = 100$$
$$X = 100$$

TABLE 10.1 Monte Carlo for European $S_0 = 100$, $X = 100$, $\sigma = 20\%$, $r = 10\%$

n	C_t	Standard Error
100	14,6543	1,8650
10,000	13,4870	0,1804
1,000,000	13,2766	0,0174
Analytical	13,2697	

$$T - t = 1$$
$$r = 10\%$$
$$\sigma = 20\%.$$

The simulation results are given in Table 10.1 and compared with the analytical European call option value. The standard error numbers are of particular interest.

The Monte Carlo method is extremely useful in pricing of contingent claims. Consult [14, 15, 17, 29, 55, 60] for further reading and insight. Higham [55] provides an excellent introduction.

CHAPTER 11

Option Values: II

The most important notion in option hedging and trading is the contamination principle: It is the fundamental principle of dynamic hedging. It means roughly that if there is a possible spot in time and space capable of bringing a profit, then the areas surrounding it need to account for that effect.

– NASSIM NICHOLAS TALEB [100]

Once the limitations of the Black-Scholes model and the subtleties of the continuous limit are fully understood, the 'blind' use of Ito calculus to obtain a first approximation to the price of derivatives becomes justified.

– BOUCHAUD AND POTTERS [14]

Better models often create nightmares. Every options trader and risk manager, while reaping the benefits of the contributions of Black, Scholes, and Merton, needs to spend his time working around the assumptions that needed to be made for the model to stand.

– NASSIM NICHOLAS TALEB [100]

Work is of two kinds: first, altering the position of matter at or near the earth's surface relatively to other such matter; second, telling other people to do so. The first kind is unpleasant and ill paid; the second is pleasant and highly paid.

- RUSSELL, BERTRAND (1872–1970)

11.1 INTRODUCTION

In Chapters 8 and 9, we showed the intuition behind the binomial method to value derivative securities. We also demonstrated that in the limit as the step-size, $\delta t \to 0$, the binomial method reduces to the Black–Scholes PDE. We discuss some of its specific solutions in Section 12.4.

In Chapter 10, we showed how to derive the value of European derivatives. In this chapter, we show how to use this specific framework to obtain the Black–Scholes PDE. We also discuss various model-based assumptions needed to achieve the result.

11.2 CONTINGENT CLAIM ANALYSIS

Let us look at the model for our stock dynamics, namely Equation (10.3), again. Using the familiar Maclaurin series expansion, we can write

$$\begin{aligned} S(t+\delta t) &= S(t) e^{(\mu - \sigma^2/2)\delta t + \sigma\sqrt{\delta t}\, z} \\ &= S(t)\left(1 + (\mu - \sigma^2/2)\delta t + \sigma\sqrt{\delta t}\, z \right. \\ &\quad \left. + \frac{1}{2}((\mu - \sigma^2/2)\delta t + \sigma\sqrt{\delta t}\, z)^2 + \cdots\right), \end{aligned}$$

hence

$$\begin{aligned} S(t+\delta t) - S(t) &= \delta S \quad\quad\quad\quad\quad\quad\quad\quad (11.1) \\ &\cong \left(\sigma\sqrt{\delta t}\, z + (\mu - \sigma^2/2 + \sigma^2/2\, z^2)\,\delta t\right) S(t). \end{aligned}$$

Incidentally, this implies that

$$E[\delta S] = \mu S \delta t,$$

and

$$E[(\delta S)^2] = \sigma^2 S^2 \delta t. \quad\quad\quad (11.2)$$

We now establish a portfolio where we hold a European call that is hedged using a position, Δ, in the underlying:

$$\Pi(S(t), t) = C(S(t), t) - \Delta S(t). \quad\quad\quad (11.3)$$

Now consider the evolution of the portfolio value over a time-step, δt, that is,

$$\delta \Pi = C(S(t+\delta t), t+\delta t) - C(S(t), t) - \Delta\, (\delta S).$$

By making use of a Taylor series expansion of $C(S(t+\delta t), t+\delta t)$, we find that

$$\delta \Pi = \partial_t C\, \delta t + \partial_S C\, \delta S + \frac{1}{2}\partial_S^2 C\, (\delta S)^2 - \Delta\, (\delta S),$$

while ignoring higher order terms and derivatives. By making the choice

$$\Delta = \partial_S C, \qquad (11.4)$$

and using Equation (11.2), we can, therefore, write

$$E[\delta \Pi] = \partial_t C\, \delta t + \frac{1}{2}\sigma^2 S^2 \partial_S^2 C\, \delta t. \qquad (11.5)$$

We would expect this portfolio to earn the risk-free rate, that is,

$$E[\delta \Pi] = r\Pi\, \delta t$$
$$= r(C - \partial_S C\, S)\delta t. \qquad (11.6)$$

Combining Equations (11.5) and (11.6), we find that

$$rC = \partial_t C + rS\partial_S C + \frac{1}{2}\sigma^2 S^2 \partial_S^2 C, \qquad (11.7)$$

after removing the δt term. There is no surprise here – we have derived the celebrated Black–Scholes PDE, see [9, 79], which governs the evolution of European derivative securities. Recall also Equation (8.25), derived as the limit of the binomial method as the size of the time-step tends to zero. We consider aspects pertaining to its solution in more detail in Chapter 12.

11.3 DISCUSSING THE HOLES IN BLACK–SCHOLES

> *Given the range of problems, it is remarkable that option formulae sometimes give values that are very close to the prices at which options trade in the market. As it stands, the Black-Scholes formula gives at least a rough approximation to the formula we would use if we knew how to take all these factors into account. Further modifications of the*

Black-Scholes formula will presumably move it in the direction of that hypothetical perfect formula.

– FISCHER BLACK (1938–1995)

Madan [74] provides the following comments:

> The Black–Merton–Scholes model, like many innovations, both succeeded and failed at the same time. It taught us that we could fly (or price options) but simultaneously created a desire for a performance level that it could not deliver.

It is, therefore, of vital practical importance to understand the assumptions we made to arrive at the Black–Scholes PDE in Equation (11.7). Let us highlight and work through these.

1. **Stock:** We assumed the underlying asset is a stock paying no dividends.
    a. Furthermore, it follows the price process given by Equation (10.3), i.e., we assume logarithmic returns are normally distributed (also referred to as the geometric Brownian motion).
       A consequence of this assumption is that the stock price moves are assumed to be continuous and jumps are not factored into the price process. Taleb [100] summarises:

       > Too bad. Given that the true world distributions do not entirely look *normal* (except by accident) we need to worry about fat tails, positive and negative skew, jumps, and other annoying matters.

       We show an interesting analysis in Section 11.6, which details the consequences of excluding jumps in a simple trading scheme.
    b. An implicit assumption throughout is that we trade without price impact, that is, our hedging activities are conducted without 'leaving a trace' in the market or affecting the stock price dynamics.
    c. We assume the stock will continue to exist, and continue to trade, for the duration of the contingent claim. Takeovers, for example, could negate this assumption.

2. **Portfolio:** We assumed that we can form the portfolio in Equation (11.3), which means:

   a. We have no preference for buying or selling (more correctly, writing) the call, C (or equivalently, a put). In practice, this will depend on risk limits typically given by various constraints on the Greeks, as defined in Chapter 15, for a typical market-maker of derivative instruments.

      Bouchaud and Protter [14] state compactly:

      > The price of an option therefore depends on the operator, on his definition of risk and his ability to hedge this risk. In the Black-Scholes model, the price is uniquely determined since all definitions of risk are equivalent (and all are zero!).

   b. We implicitly assume no credit risk in forming the portfolio. Buying an option would entail assuming some form of counterparty credit while writing the derivative would entail someone assuming our credit risk.

   c. We tacitly assume no restrictions on buying or selling S and also allow for fractional stock holdings.

   d. We assume unlimited borrowing/lending at the risk-free rate. (See the discussion in Section 11.7 involving collateral and repo markets derived in a binomial setting.)

3. **Costs:** We choose to ignore costs, to wit:

   a. We ignore costs for establishing the portfolio as described in Equation (11.3), that is, we suffer no bid or offer spreads, no brokerage, and no establishment taxes.

   b. We ignore transaction costs and taxes (e.g., capital gains tax) on further rebalancing to hedge the portfolio. This specifically pertains to the rebalancing we do by choosing the amount of stock to be held, from Equation (11.4), to remain Delta-neutral, in Equation (11.5).

      In Section 11.5, we show a relevant analysis to estimate the cost of including transaction costs while maintaining our Delta-hedging.

c. We have no margin requirements or any balance sheet restrictions (which may include specified costs and capital charges) – these could present significant hurdles. In the words of Green [53]:

> The reality is that credit, funding and capital concerns are very far from minor adjustments to the value of a single derivative contract or portfolio of derivatives.

We show the effects of including funding curves in Section 11.7 in a one-period binomial setting.

d. We do not have (nor have we experienced) forced liquidation of our positions. One such example could be a corporate takeover event forcing early termination of an option.

4. **No-arbitrage:** We assume there are no arbitrage opportunities in establishing Equation (11.6).

5. **Continuous Trading:** Our final result, Equation (11.7), contains no reference to δt. We started with an arbitrary, fixed time-step and have assumed continuous trading.

6. **Constants:** Equation (11.7) has r and σ constant; although, we can fairly easily remove those restrictions. Note that we also assume that we have an estimate or forecast of the volatility, σ.

The volatility, observed in markets where options trade, is called the implied volatility, that is, the volatility parameter used in the Black–Scholes option pricing model that yields the correct market price – it is useful, but not necessarily a meaningful forecast of the volatility of the underlying over the remaining time to maturity of the option. Recall the discussion in Section 5.3.

7. **Delta:** The Δ in Equation (11.4) raises some eyebrows; it is a function of S and t, but it is not a constant. Paul Wilmott [103] has the following synoptic explanation:

> If you hedge discretely, as you must, then Black-Scholes only works on average. But as you hedge more and more frequently, going to the limit $\delta t = 0$, then the total hedging error tends to zero, thus justifying the Black-Scholes model.

It is, therefore, instructive to consider the hedging error, that is,

$$\delta\Pi - r\Pi\delta t,$$

which, following some algebraic manipulation, we can write as

$$\frac{1}{2}\partial_S^2 C\left[(\delta S)^2 - \sigma^2 S^2 \delta t\right] \approx \frac{1}{2}\sigma^2 \delta t S^2 \partial_S^2 C\left[z^2 - 1\right],$$

the expectation of which is zero. The hedging error has a χ^2 distribution with a single degree of freedom. See, [46], for example, for details of the χ^2 distribution.

Our message should be simple: that is, use models actively, but understand their shortcomings. Emanuel Derman [37] offers compendious advice:

> Models are only models, toylike descriptions of idealized worlds. Simple models envisage a simple future; more sophisticated models incorporate a more complex set of future scenarios that can more closely approximate actual markets. But no mathematical model can capture the intricacies of human psychology. Watching traders occassionally put too much faith in the power and formalism of mathematics, I saw that if you listen to the models' siren song for too long, you may end up on the rocks or in the whirlpool.

11.4 ADJUSTMENT FOR CONTINUOUS DIVIDENDS

While the original Black–Scholes analysis [9] ignored dividends, Merton [79] demonstrated the inclusion of continuous payment of dividends, among other aspects. Following the analysis in Section 11.2, we can incorporate a continuous dividend payment.

In fact, let us assume that the underlying stock or index pays dividends at a constant, continuously compounded rate of d. Hence, we need to add the amount

$$-dS\Delta\delta t$$

to $\delta\Pi$ to account for the fact that we are short of the underlying and we will need to bear the cost of paying the dividends. The analysis proceeds directly as before, where we now obtain the so-called Black–Scholes–Merton PDE:

$$rC = \partial_t C + (r - d)S\partial_S C + \frac{1}{2}\sigma^2 S^2 \partial_S^2 C. \tag{11.8}$$

11.5 INVOKING TRANSACTION COSTS

Let us briefly recapture the essence of replicating a contingent claim as explained in Section 11.2. We form a portfolio consisting of the claim, namely Δ of the underlying asset, S, cash, and, based on moves in the underlying, re-hedge the portfolio. In our analysis, one of the assumptions has been continuous re-balancing of the portfolio.

In practice, we have costs. These could be bid-offer spreads or transaction-based charges, such as brokerage costs typically levied on the exchange of assets. We follow the analysis of Hoggard, et al. [56] to detail the impact transaction costs could have on the replication of a contingent claim.

In essence, we are trying to understand how changes in time, and S, have an impact on the value of the portfolio. In essence, what is the change in Δ, that is,

$$\partial_S C(S+\delta S, t+\delta t) - \partial_S C(S,t)?$$

We appeal to Taylor's theorem again to achieve an estimate. Using the δS defined in Equation (11.1), we note to order $\sqrt{\delta t}$ that

$$\partial_S C(S+\delta S, t+\delta t) - \partial_S C(S,t) = \partial^2_{St} C \delta t + \partial^2_S C \delta S + \cdots$$
$$\approx \partial^2_S C \sigma \sqrt{\delta t} z S.$$

Denoting the transaction cost by κ, we can, therefore, estimate the cost impact of our re-balancing activities as

$$\kappa S \left| \partial_S C(S+\delta S, t+\delta t) - \partial_S C(S,t) \right| \approx \kappa \sigma \sqrt{\delta t} S^2 \left| \partial^2_S C z \right|,$$

the expectation of which is

$$\sqrt{\frac{2 \delta t}{\pi}} \kappa \sigma S^2 \left| \partial^2_S C \right|. \tag{11.9}$$

Note the factor $\sqrt{\frac{2}{\pi}}$ arises from calculating $E[|z|]$. We can, therefore, subtract this cost from Equation (11.5), that is,

$$E[\delta \Pi] = \partial_t C \, \delta t + \frac{1}{2} \sigma^2 S^2 \partial^2_S C \, \delta t - \sqrt{\frac{2 \delta t}{\pi}} \kappa \sigma S^2 \left| \partial^2_S C \right|$$
$$= r(C - \partial_S C \, S) \, \delta t.$$

Upon dividing by δt, we have the following equation for the value of an option in the presence of transaction costs [56, 105]:

$$\partial_t C + rS\partial_S C + \frac{1}{2}\sigma^2 S^2 \partial_S^2 C - \sqrt{\frac{2}{\pi \delta t}}\kappa \sigma S^2 \left|\partial_S^2 C\right| = rC. \qquad (11.10)$$

11.6 CONSTANT PROPORTION PORTFOLIO INSURANCE

Constant proportion portfolio insurance, also known as CPPI, is a strategy invoked by investors, which allows them to limit downside risk while retaining upside potential. Further, CPPI maintains the exposure to risky assets equal to a constant multiple of the difference between the current portfolio value and a specified guaranteed amount.

In diffusive models with continuous trading, as we detail below, the strategy has no downside risk. In practice, however, the risk is not negligible due to jumps and would grow with the leverage applied to the difference between the portfolio value and the guaranteed amount.

In our Black–Scholes paradigm, we consider a portfolio consisting of a ZCB, $B(t, T)$, and a stock, $S(t)$, and work with a constant interest rate, r, for the sake of simplicity. Our strategy is denoted by $V(t)$. We want to ensure a minimum percentage, notably M, of our principal, is preserved at the maturity, T. This amount is clearly worth $MB(t, T)$ at time t.

The CPPI strategy works as follows: If we suppose $V(t) > MB(t, T)$, we hold an amount of $S(t)$ equal to

$$\alpha H(t) := \alpha \left(V(t) - MB(t,T)\right),$$

for some constant multiplier, α. Alternatively, the full strategy is invested in $B(t, T)$.

Now note that

$$\delta V = \alpha H \left(\delta S / S\right) + (V - \alpha H) r \delta t,$$

with δS defined in Equation (11.1). Furthermore, we have that

$$\delta H = \delta V - MrB\delta t.$$

A healthy dose of algebra yields:

$$\delta H = rH\delta t + \alpha H \left((\mu - \sigma^2/2 + \sigma^2/2z^2) - r\right)\delta t + \alpha H \sigma \sqrt{\delta t}z.$$

A similar analysis to the one concluded in Section 10.2 now reveals that:

$$H(T) = H(0)\exp\left(rT + \alpha(\mu - r)T - \frac{1}{2}\alpha^2\sigma^2 T + \alpha\sigma\sqrt{T}z\right).$$

It, therefore, follows that

$$V(T) = M + (1 - MB(0,T)) \quad (11.11)$$
$$\times \exp\left(rT + \alpha(\mu - r)T - \frac{1}{2}\alpha^2\sigma^2 T + \alpha\sigma\sqrt{T}z\right).$$

Equation (11.11) details quite clearly that there is no risk of the floor being breached within the Black–Scholes specified model – regardless of the multiplier value. The expected return of the CPPI strategy is given by

$$E[V(T)] = M + \left(1 - Me^{-rT}\right)\exp\left(rT + \alpha(\mu - r)T\right),$$

which, see Cont and Tankov [32], leads to the conclusion that the expected return of the CPPI strategy can be increased indefinitely by taking a high enough multiplier whenever $\mu > r$. Interesting?!

Quoting Cont and Tankov [32]:

> In reality, the possibility of reaching the floor, known as "gap risk", is widely recognised by CPPI managers: there is a nonzero probability that, during a sudden downside move, the fund manager will not have time to readjust the portfolio, which then goes crashing through the floor.

The CPPI example shows us that we can attain erroneous results because our initial model assumptions are incomplete. In this case, we assumed that logarithmic stock returns are normally distributed, which precludes jumps; this is a flawed assumption in practice.

11.7 INCLUSION OF COLLATERAL IN THE MODEL

Piterbarg [88] demonstrated how the posting of collateral affects the price of contingent claims. In this section, we follow the ideas of Hunzinger and Labuschagne [59] to implement this work in a binomial setting.

We introduce the following notation, see [59, 88]:

1. **Collateral:** r_C denotes the rate paid on collateral. The amount in the collateral account posted with the counterpart is denoted by γ_c.

2. **Repurchase:** r_R denotes the rate paid on a repurchase agreement involving the underlying stock. The amount in the repurchase account is denoted by γ_R. The account reflects the cash amount invested/borrowed in order to fund the stock position, ΔS, by use of a repurchase agreement.

3. **Funding:** r_F denotes the unsecured funding rate. The amount in the funding account is denoted by γ_F. This amount reflects the difference between the price of the hedging portfolio, γ_P, and the collateral amount, that is,

$$\gamma_F \equiv \gamma_P - \gamma_C. \tag{11.12}$$

4. **Rates:** Intuitively, it seems entirely plausible that we have

$$r_C \leq r_R \leq r_F.$$

All the rates are assumed to be deterministic and continuously compounded.

Our analysis now follows the exposition in Section 8.3, where we closely provided the derivation of the binomial model.

$$C_u = \Delta u S - \gamma_R e^{r_R \delta t} + \gamma_F e^{r_F \delta t} + \gamma_C e^{r_C \delta t} \tag{11.13}$$
$$C_d = \Delta d S - \gamma_R e^{r_R \delta t} + \gamma_F e^{r_F \delta t} + \gamma_C e^{r_C \delta t}. \tag{11.14}$$

Subtracting (11.14) from (11.13) yields Δ as before, namely

$$\Delta = \frac{C_u - C_d}{S(u-d)}.$$

Therefore, we can write the repurchase account as

$$\gamma_R = \Delta S = \frac{C_u - C_d}{(u-d)}.$$

Noting Equation (11.12), we can solve for γ_P. We observe from Equation (11.13) that

$$C_u = \Delta S \left(u - e^{r_R \delta t}\right) + \gamma_P e^{r_F \delta t} + \gamma_C \left(e^{r_C \delta t} - e^{r_F \delta t}\right),$$

therefore,

$$\gamma_P = e^{-r_F \delta t} \left[C_u - \Delta S \left(u - e^{r_R \delta t} \right) + \gamma_C \left(e^{r_F \delta t} - e^{r_C \delta t} \right) \right], \quad (11.15)$$

yields the value of the derivative. (Note, the price of the derivative has to be equal to the replicating portfolio to avoid any arbitrage opportunities). Defining

$$q = \frac{e^{r_R \delta t} - d}{u - d}, \quad (11.16)$$

we can now rewrite the value of the derivative to resemble our earlier recognisable form, namely Equation (8.11), to wit

$$C = e^{-r_F \delta t} \left[q C_u + (1 - q) C_d + \gamma_C \left(e^{r_F \delta t} - e^{r_C \delta t} \right) \right]. \quad (11.17)$$

Equivalently, we could write:

$$\begin{aligned} C(S(t), t) &= e^{-r_F \delta t} \left[q\, C(uS(t)) + (1 - q)\, C(dS(t)) + \gamma_C \left(e^{r_F \delta t} - e^{r_C \delta t} \right) \right] \\ &= e^{-r_F \delta t} E^q \left[C(S(t + \delta t), t + \delta t) \right] + e^{-r_F \delta t} \gamma_C \left(e^{r_F \delta t} - e^{r_C \delta t} \right) \\ &= e^{-r_F \delta t} E^q \left[C(S(t + \delta t), t + \delta t) \right] + \gamma_C \left(1 - e^{(r_C - r_F) \delta t} \right), \quad (11.18) \end{aligned}$$

where E^q now denotes the risk-neutral measure according to Equation (11.16).

11.8 FURTHER READING

Much has been written on the shortcomings of the Black–Scholes model. In addition to the references in the chapter, here are some interesting and important works:

1. **Derivation of Black–Scholes PDE:** See, for example, [58, 63, 84, 104].
2. **Jumps:** See, Merton [81], Cont and Tankov [30].
3. **Stochastic Volatility:** Gatheral [50].
4. **Lévy Processes:** Carr et al. [21], Schoutens [96].
5. **Hedging Impact:** See, [14, 89, 95, 100].
6. **Stock Process:** See, [4, 33, 39, 43].
7. **Transaction Costs:** See, [56, 67].

CHAPTER 12

Black–Scholes PDE

"Just the place for a Snark!" the Bellman cried, As he landed his crew with care; Supporting each man on the top of the tide By a finger entwined in his hair.

"Just the place for a Snark! I have said it twice: That alone should encourage the crew. Just the place for a Snark! I have said it thrice: What I tell you three times is true."

– "THE HUNTING OF THE SNARK" BY LEWIS CARROLL

Before long I expect the Swiss to design a watch with built-in option formulae. Or an electronic army knife that tells you how to use options to cut risk.

– FISCHER BLACK (1938–1995)

When asked how soon he expected to reach certain mathematical conclusions, Gauss replied that he had them long ago, all he was worrying about was how to reach them!

– RENÉ J. DUBOS

In order to solve this differential equation you look at it till a solution occurs to you.

– GEORGE POLYÁ (1887–1985)

12.1 INTRODUCTION

In Sections 8.7 and 11.2, we showed the derivation of the Black–Scholes PDE; recall Equations (8.25) and (11.7). In line with the quote above, we will provide a third derivation in this chapter.

12.2 GOVERNING PDE

For the sake of simplicity, let us assume we are working with constant interest rates; the continuously compounded interest rate will be denoted by r as before. We consider the value of a European derivative security, with general payoff $V(T)$, at time T.

By virtue of risk-neutral valuation (recall Equation (7.6) and also Equation (8.11) and the analyses preceding these), we can state that the value of $V(t)$ at time t is given by:

$$V(t) = e^{-r(T-t)} E^Q[V(T)], \qquad (12.1)$$

where E^Q denotes the expected value evaluated using the risk-neutral probability.

Now consider $\delta t > 0$. We can write $V(t)$, essentially a function of $S(t)$, as follows:

$$V(S(t+\delta t), t+\delta t) = V(t) + \delta t\, \partial_t V + \delta S\, \partial_S V + \frac{1}{2}(\delta S)^2\, \partial_S^2 V + \ldots \qquad (12.2)$$

where we denote $\delta S = S(t + \delta t) - S(t)$ as before. By rewriting Equation (12.1) and combining it with Equation (12.2), we find that

$$V(t)e^{r\delta t} = V(t) + \delta t\, \partial_t V + E^Q[\delta S]\, \partial_S V + \frac{1}{2} E^Q[(\delta S)^2]\, \partial_S^2 V + \ldots \qquad (12.3)$$

Using the risk-neutral expectation of the stock price, we know that

$$E^Q[\delta S] = rS\delta t,$$

and, furthermore,

$$E^Q[(\delta S)^2] \approx \sigma^2 S^2 \delta t.$$

Therefore, we can simplify Equation (12.3) to read

$$rV = \partial_t V + rS\partial_S V + \frac{1}{2}\sigma^2 S^2 \partial_S^2 V, \qquad (12.4)$$

by noting the Taylor series expansion of $e^{r\delta t}$ and dividing by $\delta t > 0$. Recall the previous derivations of the Black–Scholes PDE, that is, Equations (8.25) and (11.7).

12.3 INITIAL BOUNDARY-VALUE PROBLEM

Equation (12.4) is the celebrated Black–Scholes PDE that governs the behaviour of financial derivative security prices. The PDE needs to be augmented with appropriate boundary and initial conditions to provide unique solutions for appropriate derivative securities.

Consider the case of European derivatives, for example. We know that the value of a call with strike X, at expiry (i.e., $t = T$), must be

$$C(S, T) = \max\{S - X, 0\}. \qquad (12.5)$$

Furthermore, by virtue of elementary arbitrage relations, we know that the value of a call is always non-negative and if the underlying is at zero the call must be at zero as well, that is,

$$C(0, t) = 0, \qquad (12.6)$$

for the duration of the option's existence. We also know the value of a call option should be bounded by the underlying, that is,

$$\lim_{S \to \infty} C(S, t)/S = 1. \qquad (12.7)$$

Equation (12.5) relates to the initial conditions whereas Equations (12.6) and (12.7) pertain to the boundary conditions of the derivative security. We could, similarly, derive the initial and boundary conditions for a put option.

12.4 SOME INTERESTING SOLUTIONS

PDEs are generally tough to solve – it is, therefore, instructive to consider a couple of relatively easy solutions to the Black–Scholes PDE, Equation (12.4), and ponder the insights these render.

1. It should come as no surprise that the underlying, $S(t)$, is a solution to Equation (12.4), that is,

$$V(S,t) = S.$$

2. Similarly the value of a ZCB,

$$V(S,t) = \exp(-r(T-t)), \tag{12.8}$$

where T is the maturity date and r the continuously compounded interest rate, solves Equation (12.4).

3. Therefore,

$$V(S,t) = S - X\exp(-r(T-t)), \tag{12.9}$$

typically a forward contract, solves Equation (12.4) as well. In fact, linear combinations of the stock and bond will solve the PDE.

4. A direct consequence of Equations (12.8) and (12.9) is that if $V(S,t) = C(S,t)$, put–call parity implies that $P(S,t)$ solves:

$$rP = \partial_t P + rS\partial_S P + \frac{1}{2}\sigma^2 S^2 \partial_{SS}^2 P.$$

5. Consider a basic futures contract, F, where

$$F = Se^{r(T-t)}.$$

Defining $U(F,t) = V(S,T)$, we have:

$$\begin{aligned}
\partial_t V &= \partial_F U \, \partial_t F + \partial_t U = \partial_t U - rF\partial_F U \\
\partial_S V &= \partial_F U \, \partial_S F \\
\partial_S^2 V &= \partial_F^2 U \, (\partial_S F)^2.
\end{aligned}$$

Hence, using Equation (12.4), we find:

$$\begin{aligned}
\partial_t V &+ rS\partial_S V + \frac{1}{2}\sigma^2 S^2 \partial_S^2 V \\
&= \partial_t U - rF\partial_F U + rF\partial_F U + \frac{1}{2}\sigma^2 F^2 \partial_S^2 U \\
&= rU.
\end{aligned}$$

So, $U(F,t)$ solves the following PDE:

$$rU = \partial_t U + \frac{1}{2}\sigma^2 F^2 \partial_{FF}^2 U. \tag{12.10}$$

This example is extremely useful because many options have a corresponding futures contract as their underlying contract instead of a spot (or cash) instrument.

We detailed the derivation of the Black-76 call option prices in Equations (10.16) and (10.17). These solve the PDE in Equation (12.10).

6. If we have that $U(F,t)$ solves Equation (12.10), we can show by differentiation that

$$\frac{F}{B} U\left(\frac{B^2}{F}, t\right) \tag{12.11}$$

is a solution, as well, for any constant $B > 0$.

7. Using Equation (12.11), we consider a European put option,

$$\frac{F}{X} P\left(\frac{X^2}{F}, X, t\right),$$

at option expiration, that is,

$$\frac{F}{X} P\left(\frac{X^2}{F}, X, T\right) = \frac{F}{X}\left(X - \frac{X^2}{F}\right)^+$$
$$= \max\{F - X, 0\}.$$

Hence, we can state, after discounting, that

$$C(F, X, t) = \frac{F}{X} P\left(\frac{X^2}{F}, X, t\right). \tag{12.12}$$

This lovely result is an example of put-call symmetry.

8. Suppose we have an option with payoff $V(S,T) = S^2$. Guessing a solution of the form

$$V(S,t) = \alpha(t)S^2 + \beta(t)S + \delta(t),$$

yields

$$(\alpha'(t) + (r+\sigma^2)\alpha(t))\,S^2 + \beta'(t)S + (\delta'(t) - r\delta(t)) = 0,$$

upon substitution into Equation (12.4). From the terminal condition, we note that

$$\alpha(T) = 1, \beta(T) = \delta(T) = 0.$$

Hence, we can solve for

$$\alpha(t) = e^{(r+\sigma^2)(T-t)}, \beta(t) = \delta(t) = 0,$$

that is,

$$V(S,t) = e^{(r+\sigma^2)(T-t)}S^2.$$

The hedging strategy, therefore, as implied by Equation (11.4), holds

$$\Delta = \partial_S V = 2S e^{(r+\sigma^2)(T-t)}$$

of the underlying, S.

9. The following example follows along the same lines, albeit a bit more complicated, but quite instructive. Suppose we have a derivative instrument with payoff $V(S,T) = S(T)^N$. Its solution is given by

$$V(S,t) = e^{(N-1)(r+\frac{1}{2}\sigma^2 N)(T-t)}S^N.$$

The solution follows by guessing $V(S,t) = h(t)S^N$ as a possible solution to Equation (12.4). Substituting this choice into the PDE and performing some algebra leaves us with:

$$\partial_t h = -\alpha h,$$

where $\alpha = (N-1)(r+\frac{1}{2}\sigma^2 N)$ and the solution follows directly.

This example illustrates the so-called Delta-hedging argument. If $N \neq 1$, the presence of σ^2 in the solution shows that we will need to rebalance the portfolio as the spot moves and we are approximating a nonlinear payoff, with a linear instrument, that is, a combination of the spot and a bond or cash holding.

10. The following example is more tedious to prove, but still provides a beautiful result. The full proof follows by induction. Suppose that $V(S, t)$ solves Equation (12.4). Then, for positive integers, N:

$$V_N(S, t) = S^N \partial_S^N V,$$

solves Equation (12.4) as well.

11. We devote Chapter 13 to the special case of perpetual options.

12.5 DUAL BLACK–SCHOLES PDE

Suppose we have European calls of all exercise prices available. Then, regarding S as fixed and X as variable, we can derive a 'dual' Black–Scholes PDE. We provide brief details below; recall the Euler relation in Equation (6.9),

$$S\partial_S C + X\partial_X C = C,$$

and Equation (6.10):

$$\partial_S^2 C = \left(\frac{X}{S}\right)^2 \partial_X^2 C.$$

If we subsitute these equations into the Black–Scholes PDE, Equation (12.4), we obtain the following equation

$$\partial_t C + r(C - X\partial_X C) + \frac{1}{2}\sigma^2 S^2 \left(\frac{X^2}{S^2}\right) \partial_S^2 C = rC.$$

Hence, a simple time transformation yields:

$$\partial_T C + rX\partial_X C - \frac{1}{2}X^2\sigma^2\partial_X^2 C = 0. \tag{12.13}$$

Equation (12.13) is frequently referred to as the dual Black–Scholes PDE. It is frequently used to define the so-called local volatility surface, which can now be written as follows

$$\sigma^2(X, T) = \left(\frac{2}{X^2}\right) \frac{\partial_T C + rX\partial_X C}{\partial_X^2 C}, \tag{12.14}$$

across a range of European option maturities and strikes.

Note, practical implementation of Equation (12.14) is difficult. We have limited option strikes available; hence, the numerical estimates of $\partial_T C, \partial_X C$ and $\partial_X^2 C$ are potentially quite inaccurate.

Derman and Miller [40] provide a very readable insightful account of local volatility. See, also Gatheral [50] and Wilmott [104].

12.6 EXAMINING A SPECIAL CASE: $\sigma = 0$

We frequently obtain insights in applied mathematics by considering special cases of governing equations. Let us illustrate this point by considering the special case where $\sigma = 0$ in Equation (12.4). First, consider the transformation

$$\tau = r(T - t).$$

We define a new function

$$V(S, t) = v(S, \tau),$$

whereupon substitution into Equation (12.4) renders a new PDE,

$$\partial_\tau v - S \partial_S v - \frac{\sigma^2}{2r} S^2 \partial_S^2 v + v = 0.$$

Also, the terminal value, Equation (12.5), has now been transformed to an initial value,

$$v(S, 0) = \max\{0, S - X\}.$$

In the special case we want to consider, namely where $\sigma = 0$, we have that

$$\partial_\tau v - S \partial_S v + v = 0.$$

We define a new function,

$$v(S, \tau) = S w(S, \tau),$$

which gives

$$\partial_\tau w - S \partial_S w = 0,$$

upon substitution. This equation can be solved using the well-known method of characteristics, the formal solution is

$$w(S, \tau) = w_0(Se^\tau),$$

where $w_0(x) = w(x, 0) = v(x, 0)/x$. In terms of our original formulation, therefore,

$$V(S, t) = e^{-r(T-t)}(S(T)e^{r(T-t)} - X)^+.$$

In words, we already know the value of an option is given by the discounted value of the expected payoff under the risk-neutral measure, by virtue of Equation (7.6). Here, given that the volatility, $\sigma = 0$, all the uncertainty is removed and we have that the call option value is the discounted value of the intrinsic value of the forward (i.e., the carry-adjusted spot). See, Sauer [94] for a complete discussion.

12.7 REDUCTION TO THE HEAT EQUATION

In this section, we will discuss the reduction of the Black–Scholes PDE to the well-known heat equation of mathematical physics. We note this novel transformation technique from Davis [35].

We make use of Equation (10.8), rewritten as

$$S(t) = S(0)e^{(r-\sigma^2/2)t+\sigma W(t)},$$

where

$$W(t) = \sqrt{t}\, z.$$

We now solve for W as

$$W := g(S, t) = (\ln(S(t)/S(0)) - (r - \sigma^2/2)t)/\sigma.$$

Calculating derivatives of $g(S, t)$, we obtain:

$$\partial_t g = \frac{-(r - \sigma^2/2)}{\sigma}$$

$$\partial_S g = \frac{1}{\sigma S}.$$

If we now view $C(S,t) = B(W,t)$, we obtain the following derivatives:

$$\begin{aligned}
\partial_t C &= \partial_t B - \partial_g B \, \partial_t g \\
\partial_S C &= \partial_g B \, \partial_S g \\
\partial_S^2 C &= \partial_g^2 B \, (\partial_S g)^2 + \partial_g B \, \partial_S^2 g.
\end{aligned}$$

Upon substituting all these equations into the Black–Scholes PDE, Equation (12.4), we obtain the in-homogeneous heat equation, namely

$$\partial_t B + \frac{1}{2} \partial_W^2 B = rB,$$

after completing some algebraic simplification. Employing a time transformation, $\tau = T - t$ and setting $A(W, \tau) = e^{r\tau} B(W, \tau)$, we find the standard heat equation, namely

$$\partial_\tau A = \frac{1}{2} \partial_W^2 A,$$

with solution, see Polyanin [90], for example,

$$A(W, \tau) = \frac{1}{\sqrt{2\pi\tau}} \int_{-\infty}^{\infty} \exp\left(-\frac{(W - \xi)^2}{2\tau}\right) f(\xi) d\xi, \qquad (12.15)$$

subject to initial condition $A(W, 0) = f(W)$. When we subsitute all the elements back, we obtain the Black–Scholes price in Equation (10.12).

12.8 FURTHER READING

Wilmot [104] and Wilmot et al. [105] provide extensive insights into the analytical solution of the Black–Scholes PDE (with extensions to exotic options). Numerical solution techniques are also covered. See, also, Kwok [65]. The notes by Prof R. Kohn are instructive, https://math.nyu.edu/~kohn/pde_finance.html.

Higham [55] provides a insightful description of the finite difference method used to solve the Black–Scholes PDE. He also shows the equivalence of the explicit finite difference method to the binomial option pricing method.

CHAPTER 13

Perpetual Options

To infinity and beyond ...

— BUZZ LIGHTYEAR, TOY STORY, WALT DISNEY-PIXAR

13.1 INTRODUCTION

We have shown the derivation of the Black–Scholes PDE and demonstrated some specific interesting solutions. In this chapter, we consider perpetual options, that is, options with an infinite lifespan.

It seems intuitively plausible that the solution of a perpetual option should be independent of time. Dependence on time would mean that the option would have an associated Theta (i.e., time decay); how can an option's value decay indefinitely, yet have a finite value?

13.2 GOVERNING DIFFERENTIAL EQUATION

We return to the Black–Scholes PDE in Equation (12.4), but now without time dependence, that is,

$$rS\frac{df}{dS} + \frac{1}{2}\sigma^2 S^2 \frac{d^2 f}{dS^2} = rf. \qquad (13.1)$$

Equation (13.1), is known as a Cauchy–Euler type differential equation. By inspection, we note that this equation, typically, admits solutions of the form S^α.

Upon substitution of our "guess S^α into" the differential equation, we have:

$$\alpha r + \frac{1}{2}\alpha(\alpha - 1)\sigma^2 = r, \tag{13.2}$$

the solutions of which are given by $\alpha = 1$ or $\alpha = -2r/\sigma^2 := -\nu$. The general solution of Equation (13.1) is, therefore, given by

$$f(S) = AS + BS^{-\nu}, \tag{13.3}$$

for some constants, A and B, to be determined from the relevant boundary conditions.

We shall now turn to some specific examples. These examples illustrate the effect of different boundary conditions as well as path dependency in the problem.

13.3 PERPETUAL EUROPEAN CALL

Let us consider a perpetual European call option. We already know that the solution will be in the form of Equation (13.3), subject to obtaining specific values for A and B. These would typically follow from the specific boundary conditions given by Equations (12.6) and (12.7).

Consider Equation (12.6). It is clear from Equation (13.3) that

$$\lim_{S \to 0} f(S) = \infty$$

for nonzero B. In order to ensure finite solutions, we have to force $B = 0$. Similarly, from Equation (12.7) we require that $A = 1$. The value of a perpetual European call is, therefore, given by

$$f(S) = S. \tag{13.4}$$

This is an interesting solution in its own right. A perpetual European option could clearly never be exercised. This begs the question why such an option would have any value?

The solution obtained is consistent with our previous work. If we consider the option bounds in Section 6.3, we observe from Equations (6.2) and (6.3) that, in the limit as $T \to \infty$, a European call must be equal to the value of the underlying.

It is worth noting that the option could be sold between counterparties (ignoring credit risk for the time being). This is an important principle worth remembering. Options are bought and sold for their gearing nature;

take equity warrants for example, these are very seldom held until expiry. A similar principle would apply to the perpetual European call option.

13.4 PERPETUAL EUROPEAN CALL WITH A DOWN-AND-OUT BARRIER

Consider a perpetual European call with a so-called, down-and-out barrier, H, chosen such that $H < X$. Should the spot, S, trade at, or below, the barrier, H, the option will cease to exist.

From Equation (12.6), we require that

$$\lim_{S \to \infty} f(S)/S = 1,$$

implying that $A = 1$. The boundary condition, Equation (12.7), is, however, replaced with

$$f(H) = 0.$$

Therefore, we see that

$$H + BH^{-\nu} = 0.$$

Solving for B and substituting in the general solution, Equation (13.3), we have the specific relevant solution

$$f(S) = S - H(S/H)^{-\nu}, \tag{13.5}$$

which holds for $S > H$. It is interesting to note the resemblance with the perpetual European call option.

Consider the Delta of the down-and-out option, for $S > H$:

$$f'(S) = 1 + \nu \left(\frac{H}{S}\right)^{\nu+1}.$$

Clearly, for $S \downarrow H$, we have that $f'(S) > 1$ and increases, but for $S < H$, we have $f'(S) = 0$. Hence, we have a discontinuous Delta and a so-called Gamma spike. In a market that moves lower, we will have to buy the underlying, but once we fall below the barrier we need to liquidate everything – this leads to a very risky situation.

13.5 PERPETUAL AMERICAN PUT

American-type options are notoriously difficult to solve. The difficulty arises from the so-called free-boundary value problem that we encountered earlier in Chapter 9. In the case of perpetual American options, however, we can sometimes solve the free boundary value problem [72, 83, 104]. In this section, we consider the solution of a perpetual American put option.

We argue that it makes sense to exercise this option as the share price decreases significantly. We shall denote the largest share price below where it makes sense to exercise the put option by G. Hence, if we use the notation P to denote the price of the American put option, we can write the following boundary conditions:

$$\lim_{S \to \infty} P(S) = 0, \qquad (13.6)$$

and

$$P(G) = X - G. \qquad (13.7)$$

We note that our general solution will again be of the form (13.3), namely

$$f(S) = AS + BS^{-\nu}.$$

Using condition (13.6), we see that we need to choose $A = 0$ in order to have finite solutions. From condition (13.7), we, therefore, find that

$$P(G) = X - G = BG^{-\nu},$$

which means that

$$B = (X - G)G^{\nu}.$$

We can, therefore, express the solution of the American put options as:

$$P(S) = (X - G)(S/G)^{-\nu}, \qquad (13.8)$$

subject to finding the value for G. To find the value of G, we calculate the derivative of P in Equation (13.8) with respect to G and solve for the specific G value for which the derivative would be equal to zero. We find that this value is given by

$$G = \frac{\nu X}{1+\nu}. \tag{13.9}$$

This is the free-boundary value for the American put. Substituting this value into Equation (13.8) gives us the solution

$$P(S) = \left(\frac{X}{1+\nu}\right)\left(\frac{(1+\nu)S}{\nu X}\right)^{-\nu}. \tag{13.10}$$

13.6 COMMENTS AND FURTHER READING

We discussed the specific case of perpetual options on an underlying that pays no dividends. This was done to illustrate the concept and methodology. Wilmott [104], for example, considers perpetual puts and calls on an underlying paying continuous dividends.

Wilmott *et al.* [105] show how to solve two interesting perpetual options. The so-called Russian option is a perpetual option paying the maximum realised asset price at a time chosen by the holder. The stop-loss option can be thought of as a perpetual lookback barrier with a rebate that is a fixed proportion of the maximum realised value of the asset price.

CHAPTER 14

Application: Corporate Credit

Remember that credit is money.

– BENJAMIN FRANKLIN (1705–1790)

Someone stole all my credit cards, but I won't be reporting it. The thief spends less than my wife did.

– HENNY YOUNGMAN

How did you go bankrupt? Two ways: Gradually and then suddenly.

– ERNEST HEMMINGWAY (1899–1961)

14.1 INTRODUCTION

In practice, we are faced with contingent claims that might be implied, imputed, or embedded. As an example of the application of the theory of contingent claims, in 1974, Merton [80] suggested an option theoretic framework for modelling corporate default. His basic proposal entailed modelling default as the event that occurs when the value of the assets of a firm falls below the value of the firm's debt.

In the rest of this chapter, we draw on the work of Kane [86].

14.2 CAPITAL STRUCTURE OF THE FIRM

We assume that the following holds:

1. **Debt:** The firm is funded by zero-coupon debt, which expires at time, T, with face value, F.

2. **Equity:** The firm has equity, E, which pays no dividend.

In standard accounting terms, we have that the value of the assets of the firm, $A(t)$, is exactly equivalent to the sum of debt, $D(t)$, and equity, $E(t)$, that is,

$$A(t) \equiv E(t) + D(t).$$

We assume that default can only happen at the time the debt matures – we therefore have two outcomes at time, T:

1. **Solvency:** If $A(T) \geq F$, the firm is solvent and the equity owners retain

$$A(T) - F,$$

    after repaying debt.

2. **Insolvency:** If $A(T) < F$, the firm is insolvent and debt holders will liquidate $A(T)$ and shareholders are left with no value.

Therefore, we have that

$$E(T) = \max\{A(T) - F, 0\}, \tag{14.1}$$

that is, we can view equity as a call option on the assets of a firm with a strike at the face value of the debt.

We can therefore use the Black–Scholes–Merton option pricing formula to calculate the value of $E(t)$. For the sake of simplicity, we assume that interest rates are constant. Furthermore, we assume that $A(t)$ evolves according to Equation (10.4), as derived in Section 10.2, that is,

$$A(T) = A(t)e^{(\mu - \sigma_A^2/2)(T-t) + \sigma_A \sqrt{T-t}\, z(T)},$$

where σ_A denotes the volatility of the assets and $\mu = r$, typically, in the risk-neutral measure. Furthermore $Z(T) \sim N(0,1)$. As before, using Equation (7.6), we can evaluate $E(t)$ as:

$$E(t) = A(t)N(d_1) - Fe^{-r(T-t)}N(d_2), \tag{14.2}$$

where

$$d_1 = \frac{\ln(A(t)/F) + (r + \sigma_A^2/2)(T-t)}{\sigma_A\sqrt{T-t}},$$

and

$$d_2 = d_1 - \sigma_A\sqrt{T-t}.$$

We can write the payoff for debt holders as:

$$D(T) = \min\{F, A(T)\}, \tag{14.3}$$

by virtue of our reasoning. Equivalently, we write Equation (14.3) as:

$$D(T) = F - \max\{0, F - A(T)\}, \tag{14.4}$$

which details the logic that the value of the debt is equivalent to being long of a risk-free bond (ZCB) and short of a put option on the assets of the firm struck at the face value of the debt.

Using put–call parity, we rewrite Equation (14.4) as:

$$D(T) = A(T) - \max\{A(T) - F, 0\}, \tag{14.5}$$

which means that the debt value can be written as:

$$D(t) = A(t)N(-d_1) - Fe^{-r(T-t)}N(d_2). \tag{14.6}$$

We normally define $D(t)$ in terms of the continuously compounded risk-free rate and credit spread, s, namely

$$D(t) = Fe^{-(r+s)(T-t)}, \tag{14.7}$$

when interest rates and credit spreads are assumed constant. Hence, in terms of Equation (14.6), we have

$$s = -\frac{1}{(T-t)} \ln\left(\frac{D(t,T)}{F}\right) - r.$$

We can extend the formula for interest rates that are not constant.

14.3 RECOVERY

The probability that the asset value of the firm is above the face value of the debt on the maturity date is referred to as the survival probability, which we can calculate as:

$$\begin{aligned} Q(t,T) &= P(A(T) \geq F) \\ &= N(d_2). \end{aligned}$$

In the event of a default, the debt holders would want to liquidate the remaining assets of the firm. We can calculate the expected recovery value R of the debt upon default as well. We note that $D(t)$ should equate the present value of the face value times the sum of the survival probability plus the recovery percentage times the probability of not surviving, that is, at time, t,

$$D(t) = Fe^{-r(T-t)}(Q(t,T) + (1-Q(t,T))R).$$

This means that R can be computed by making use of Equation (14.6) as:

$$R = \frac{A(t)N(-d_1)}{Fe^{-r(T-t)}N(-d_2)}.$$

This is a neat extension of the model – not only are we able to price equity and bonds within the Merton framework, but we can also calculate the debt recovery value endogenously.

14.4 PRACTICAL APPLICATION AND LIMITATIONS

Merton's model [80] provides an attractive use of the Black–Scholes [9] and Merton [79] option pricing methodology to analyse corporate default and basic corporate capital structure.

We can extend the method to the valuation of more complex debt structures such as a combination of senior and subordinated debt. To illustrate the concept simply, suppose we have two debt issues (both maturing at time, T) with face value, F_s, for the senior tranche and face value, F_j, for the junior, subordinated tranche. Default will occur if

$$A(T) < F = F_s + F_j.$$

Hence,

$$D_s(T) = F - \max\{0, A(T) - F_s\}$$
$$D_j(T) = \max\{0, A(T) - F_s\} - \max\{0, A(T) - F\}.$$

Clearly, total debt will be $D_s(T) + D_j(T) = F - \max\{0, A(T) - F\}$ and we will invoke the analysis starting with Equation (14.6).

Practical applications, however, remain constrained. Some of the reasons are:

1. **Capital Structure:** The simplified capital structure assumption is unrealistic. Adding more levels of detail is possible but analytically cumbersome [86].

2. **Default:** The model only allows a single default at time T. Bonds with coupons or different maturities become complicated, analytically, as the coupon bond is seen as a compound option. See Geske [51].

3. **Transparency:** There is generally limited transparency regarding the value of the assets of the company. Statutory reports for listed entities are issued periodically (three-to-six-monthly) meaning that an up-to-date valuation is almost impossible.

4. **Spreads:** It is easy to show that the credit spread tends to zero within this model as time tends to zero if $A(t) > F$. This is unrealistic; in practice, we have clearly differentiated non-zero spreads even for short-maturity debt.

As discussed above, one drawback of Merton's model is that we need the asset value, $A(t)$, as well as the asset volatility, σ_A, as inputs. Both parameters are not directly observable or easily available in practice.

However, the equity value, $E(t)$, as well as equity volatility, σ_E, are frequently directly observable. In many markets where we have listed equity, we would find some derivative trading on single names, especially the larger market-capitalisation based shares. (We would also be able to derive an equity volatility from the analysis of historical volatility, for example.)

However, we can use Equation (14.2) as well as invoke Ito's lemma to show that

$$\sigma_E = \sigma_A N(d_1) \frac{A(t)}{E(t)}, \tag{14.8}$$

by using Equations (14.2) and (14.8) we note that we have two equations, with two unknowns, to solve for $A(t)$ as well as σ_A.

CHAPTER 15

Greeks

Delta is the king, gamma is the queen, theta is the jester, and vega is the fortune teller.

– ANONYMOUS

In a two-dimensional approach (i.e. looking at P&L distribution at expiry) most of the Greeks are disregarded, while during the lifetime of the option they can make or break the strategy.

– P. URSONE

To make the mathematical continuous-time finance delta relevant, it needs to be accompanied by the second derivative, the gamma, and at least a third one, the DgammaDspot. In addition, because volatility moves when markets move, adding the vega to the exposure would be necessary.

– NASSIM NICHOLAS TALEB [100]

Anyone who isn't confused doesn't really understand the situation.

– EDWARD R. MURROW (1908–1965)

15.1 INTRODUCTION

In Section 12.4, we derived the Black–Scholes PDE. The derivation is subject to a number of assumptions, which we discussed in Section 11.3. One of the assumptions is that we are able to trade a certain amount of the underlying, the so-called Delta, continuously. Joshi [63] writes:

> The model requires the option seller to hedge his exposure by holding $\frac{\partial O}{\partial S}$ units of the the stock, S, at any time. This quantity is known as the *Delta* of the option and the hedging strategy is known as *Delta-hedging*.

He continues [63]:

> If we allow ourselves to use options to hedge, we can do better. The second derivative can also be matched and the portfolio's change in value for small changes in price of the underlying will now be proportional to the cube of the change, which is much smaller again. The second derivative with respect to spot is called the *Gamma* and the process we have discussed is called *Gamma-hedging*.

The *Delta* and the *Gamma* are examples of the so-called option 'Greeks' that we use to determine the exposure of financial derivatives. Consider a derivative security (or contingent claim) with price $F = F(S, X, t, T, r, \sigma)$, recall, for example, the Black–Scholes value of a European call option (see, Equation (15.2) for convenience). So, using Taylor's theorem, we could state:

$$\begin{aligned} \delta F &= F(S + \delta S, X, t + \delta t, T, r + \delta r, \sigma + \delta \sigma) - F(S, X, t, T, r, \sigma) \\ &= \partial_S F \delta S + \partial_t F \delta t + \partial_r F \delta r + \partial_\sigma F \delta \sigma + ... \end{aligned} \quad (15.1)$$

where we have only shown the first-order terms explicitly. We will discuss these and higher order terms below. The contingent claim, therefore, has multiple sensitivities to consider.

The option 'Greeks' denote the sensitivity of the price of a derivative security with respect to moves of the underlying, variables, or parameters. The Greeks are central to the hedging of contingent claims. The Greeks are calculated by differentiation of pricing formulas with respect to the underlying or parameters. Greeks can be calculated analytically for closed-form solutions of the derivative security but the calculations are quite cumbersome; hence, practically, they are frequently evaluated numerically. We return to this practical aspect in Section 15.5.

We list some of the most common Greeks below with a short description of their practical use or importance. We continue using the derivative security denoted by F in what follows.

1. **First-Order Greeks:**

    a. **Delta:** The rate of change of the derivative with respect to the underlying asset, that is,
    $$\Delta := \partial_S F \equiv \frac{\partial F}{\partial S}.$$
    See Equation (15.6) for the Delta of a European call option. The Delta of a put option can easily be derived using put/call parity.

    b. **Theta:** The rate of change of the derivative price with respect to time, Θ, is given by
    $$\Theta := \partial_t F \equiv \frac{\partial F}{\partial t}.$$

    c. **Volatility and Variance:** We can define the derivative sensitivity to volatility or variance of the underlying asset.

    Sensitivity of the derivative, with respect to the volatility σ, is given by the so-called Vega, that is,
    $$V := \partial_\sigma F \equiv \frac{\partial F}{\partial \sigma}.$$
    See Equation (15.7) for the Vega of a European option.

    Sensitivity of the derivative, with respect to the variance, σ^2, is given by the so-called Kappa, that is,
    $$\kappa := \partial_{\sigma^2} F \equiv \frac{\partial F}{\partial \sigma^2}.$$
    Note, our choice of notation above follows Derman and Miller [40]; they also note that this notation is reversed by some authors.

    d. **Rho:** The ρ describes sensitivity with respect to rates, that is,
    $$\rho := \partial_r F \equiv \frac{\partial F}{\partial r}.$$
    Refer to Equation (15.8) for Rho of a European call option.

Note that we could have multiple Rho calculations for an option. If we consider an index, which yields a continuous dividend yield, we might want to calculate the sensitivity of the option with respect to the dividend yield. Another example could be a foreign interest rate.

2. **Second-Order Greeks:**

   a. **Gamma:** The sensitivity of the Delta to the underlying asset, Γ, is described by

   $$\Gamma := \partial_S^2 F \equiv \frac{\partial^2 F}{\partial S^2}.$$

   b. **Charm:** The sensitivity of the Delta to changes in time is described as the Charm, that is,

   $$\partial_{tS}^2 F \equiv \frac{\partial^2 F}{\partial S \partial t}.$$

   c. **Volga:** Also known as 'Vega of Vega', the Volga is given by

   $$\partial_{\sigma\sigma}^2 F \equiv \frac{\partial^2 F}{\partial \sigma^2}.$$

   Also referred to as Vomma (or 'vol Gamma').

   d. **Vanna:** Vega of Delta, that is,

   $$\partial_{S\sigma}^2 F \equiv \frac{\partial^2 F}{\partial S \partial \sigma}.$$

3. **Third-Order Greeks:**

   a. **Colour:** Colour (or 'Gamma-bleed') describes how Γ moves with time, that is,

   $$\partial_{SSt}^3 F \equiv \frac{\partial^3 F}{\partial^2 S \partial t}.$$

   b. **Speed:** Changes in Gamma in relation to change in the underlying is called speed (or 'DgammaDspot'), that is,

   $$\partial_{SSS}^3 F \equiv \frac{\partial^3 F}{\partial^3 S}.$$

c. **Zomma:** Changes in Gamma in relation to changes in volatility (or 'DgammaDvol'), that is,

$$\partial^3_{SS\sigma} F \equiv \frac{\partial^3 F}{\partial^2 S \partial \sigma}.$$

Note, by virtue of the definitions above, that we can rewrite the Black–Scholes PDE, Equation (12.4), as follows:

$$\Theta + rS\Delta + \frac{1}{2}\sigma^2 S^2 \Gamma = rF.$$

This is a useful interpretation of the PDE in the context of a portfolio. In essence, if we were Delta-neutral, we see that Θ is effectively the price we pay for Γ.

15.2 BASIC FORMULAE

We repeat the Black–Scholes formula for a European call option for ease of reference purposes. Consider a call option on a stock that pays no dividends; the fair Black–Scholes option price, that is, the solution to the PDE in Equation (12.4), is given by:

$$C = SN(h^+) - Xe^{-rT} N(h^-), \qquad (15.2)$$

where $h^\pm$ is given by:

$$h^\pm = \left[\ln(\frac{Se^{rT}}{X}) \pm \frac{\sigma^2}{2}T\right] / (\sigma\sqrt{T}),$$

subject to the agreed/known parameters S, X, r, σ, and T. Furthermore, the cumulative Normal distribution function is defined as

$$N(x) = \frac{1}{\sqrt{2\pi}} \int_{-\infty}^{x} e^{-s^2/2} ds.$$

15.3 DERIVING SENSITIVITIES

We would, for theoretical purposes, typically use the formula in Equation (15.2) not only to calculate option values but also use derivatives thereof to calculate sensitivities of the option price with respect to moves of the underlying variables (the 'Greeks'). An elegant way of deriving option sensitivities is based on the following technique (attributed to Lane Hughston [57]).

We take the derivative of the call option in Equation (15.2) with respect to an arbitrary variable, y. Hence,

$$\partial_y C = \partial_y[SN(h^+)] - \partial_y[Xe^{-rT}N(h^-)],$$

which we can expand to read

$$\begin{aligned}\partial_y C &= (\partial_y S)N(h^+) - (\partial_y(Xe^{-rT}))N(h^-) \\ &\quad + S\partial_y N(h^+) - Xe^{-rT}\partial_y N(h^-).\end{aligned} \quad (15.3)$$

Note, now, the important part where we evaluate derivatives of $N(\cdot)$, namely

$$\partial_y N(h^\pm) = \frac{1}{\sqrt{2\pi}} e^{-(h^\pm)^2/2} \partial_y h^\pm. \quad (15.4)$$

We need a bit of algebra to continue evaluating the above expression. Note that

$$(h^\pm)^2 = \frac{1}{\sigma^2 T}\left[\left(\ln\left(\frac{Se^{rT}}{X}\right)\right)^2 + \frac{1}{4}\sigma^4 T^2 \pm \sigma^2 T \ln\left(\frac{Se^{rT}}{X}\right)\right].$$

Hence,

$$e^{-(h^\pm)^2/2} = e^{-\zeta}\left(\frac{Se^{rT}}{X}\right)^{\mp\frac{1}{2}}$$

where we have defined

$$\zeta = \frac{1}{2\sigma^2 T}\left[\left(\ln(\frac{Se^{rT}}{X})\right)^2 + \frac{1}{4}\sigma^4 T^2\right].$$

We now combine Equation (15.3) with the result in Equation (15.4). Therefore, we find that the sensitivity of the option price with respect to the dummy variable, y, is given by:

$$\begin{aligned}\partial_y C &= (\partial_y S)\, N(h^+) - (\partial_y(Xe^{-rT}))\, N(h^-) \\ &\quad + \frac{1}{\sqrt{2\pi}}\left(\frac{SX}{e^{rT}}\right)^{\frac{1}{2}} e^{-\zeta}\partial_y(\sigma\sqrt{T}). \end{aligned} \quad (15.5)$$

15.4 BASIC GREEKS

Let us discuss some applications of Equation (15.5).

1. **Delta:** Take the Delta of an option $\Delta = \partial_S C$. We now substitute $y = S$ in Equation (15.5) yielding

$$\partial_S C = N(h^+). \qquad (15.6)$$

The Delta Δ of an option forms a significant part of risk management of derivative profiles. In Table 15.1, we show the Delta evolution for a European option with strike $X = 85$ and various times to maturity using volatility $\sigma = 20\%$ and $r = d = 0\%$ for illustrative purposes.

2. **Vega:** Similar to the calculation of the Delta, the Vega of the option, $\partial_\sigma C$, is given by

$$\partial_\sigma C = \frac{1}{\sqrt{2\pi}} \left(\frac{SX}{e^{rT}}\right)^{\frac{1}{2}} e^{-\zeta} \sqrt{T}. \qquad (15.7)$$

3. **Rho:** Interest rate sensitivity is given by the Rho of the option, namely

$$\partial_r C = XTe^{-rT} N(h^-). \qquad (15.8)$$

4. **Exotics:** Note that Equation (15.5) was derived for the Black–Scholes price of vanilla European options. We would be able to derive some

TABLE 15.1 European Call Δ: $X = 85, \sigma = 20\%, r = d = 0\%$

Spot/Maturity	0.25	0.5	1
100.0	0.95	0.89	0.82
97.5	0.92	0.85	0.78
95.0	0.88	0.80	0.74
92.5	0.81	0.75	0.70
90.0	0.73	0.68	0.65
87.5	0.63	0.61	0.60
85.0	0.52	0.53	0.54
82.5	0.40	0.44	0.48
80.0	0.29	0.36	0.42
77.5	0.19	0.28	0.36
75.0	0.11	0.21	0.30
72.5	0.06	0.15	0.24
70.0	0.03	0.10	0.19

exotic derivatives Greeks but it is wise to heed the caution by De Weert [41], namely

> ..., exotic options have certain trigger points where the Greeks 'blow up' and therefore are not smooth at all.

15.5 PRACTICAL ASPECTS

Let us now turn to a practical setting. We will typically, in practice, run a so-called 'book' consisting of multiple option strikes and multiple maturities – even in a vanilla-only derivatives setting.

In practice, therefore, when we hedge a portfolio of derivative securities, we need to carry awareness of all the different variables that could affect (or influence) the value of the portfolio as well as our potential exposures.

In this regard, it is useful to take note of Taleb's comment [100]:

> Continuous time models should be used for pricing and getting a benchmark fair value, not to hedge.

In our hedging programme, we use the so-called Greeks described above to measure these sensitivities, but we need to adjust our approach. Taleb [100] explains the Delta, for example:

> ... a delta depends on the operator's perception of future volatility and his utility, as well as his possible frequency of adjustments.

Practically, we need to take into account that we are operating in a dynamic setting where time changes; we work off a term-structure of rates, and we interpolate from a volatility surface to obtain our derivative security inputs. We can, therefore, not use the mathematical continuous-time sensitivities, but need to resort to numerical calculations. Hence, we require full revaluation of all expressions across curves and surfaces. This means that Greek evaluation is typically performed adequately by making use of numerical differentiation.

We derived the Greeks above by assuming that a specific variable moves with the others being fixed, which is the ordinary definition of a partial derivative. It is practically also important to understand how co-movement of variables will occur. As an example, a fall in the value of a stock index would typically be associated with an increase in volatility. If we, for example, owned an out-of-the-money put option, which was traded Delta-neutral, we might find ourselves positioned short of the market as a function of the volatility increase as the market falls. Thus, we need to consider the sensitivity of the derivative with respect to both these variables. This is typically referred to as a 'shadow-Greek' exposure [100].

In conclusion, risk management of derivative products forms an integral part of the whole science of Financial Engineering – the nonlinear nature of financial derivatives requires utmost management care. Most handbooks on financial derivatives will contain some description of the Greeks. Higham [55], for example, demonstrates detailed Delta-hedging examples. Taleb [100] provides deep practical insights. See, also, [14, 25, 28, 40, 41, 92, 93, 105].

CHAPTER 16

Exotic Derivatives

We should probably stop trading derivatives, anything more complex than regular options ... I am an options trader, and I don't understand options.

– NASSIM NICHOLAS TALEB

Why, a four-year-old child could understand this contract. Run out and find me a four-year-old child. I can't make head nor tail out of it.

– GROUCHO MARX (1890–1977)

We think of the dinosaur as a byword for something past its sell-by date, but they lasted 165 million years. I hope we are that successful.

– DOUGLAS ADAMS (1952–2001)

If you can keep your head when all about you are losing theirs, you probably don't understand the problem.

– JEAN KERR

16.1 INTRODUCTION

Vanilla European derivatives afford us the ability to profit from a correct view of the distribution of an underlying asset's return at the maturity date. We frequently desire to take account of payoff profiles where some aspect of the way in which the underlying has moved is taken into account.

Emanuel Derman [37] provides us with insights:

> Behind all of these exotic structures, if they were successful in the marketplace, were two obvious principles. First, since an option is a type of bet on a future scenario that may never occur, investors want to pay as little as possible for it. Second, in order to minimize an option's cost, you must define as precisely as possible the exact scenario you are betting on. The more precise you can be about the scenario from which you want to benefit or protect yourself, the less you pay.

We shall now consider different exotic options as well as a 'classification' of these options.

16.2 WHAT DO I GET, AND WHEN?

Options represent contingent rights, with no associated obligations. A contingent right specifies a certain benefit, or payoff (and how it is determined), and when it is due. Let us examine this statement:

1. **Maturity:** When do I get it?
   It is important to note that various options exist that pertain to the time when exercise of the option could take place. Typically:

   a. **European Options:** These rights are exercised only at maturity.

   b. **American Options:** These rights can be exercised before or at maturity.

   c. **Bermudan Options:** These rights are exercised at predetermined dates and at maturity.

   d. **Trigger Options:** The contingent claim is terminated based on some market (or external) event.

2. **Payoff:** What do I get?
   Most exotic options have been designed to benefit from some level of path dependence, that is, the payoff of the option does not only depend on the terminal underlying spot level but on the way it reaches that level. We distinguish between *weak path dependence* and *strong path dependence*.

Options whose value depends on the asset history but whose valuation can still be written as a function of the spot price are termed weakly path dependent. A typical example would be a barrier option.

Strongly path dependent options depend not only on the current spot level but also on some property of its historical performance. An example would be an Asian option whose valuation depends on the average of the spot levels over a period of time.

In the payoff formulae below, the payoff could be a function of the terminal spot value, S_T, only, for example, or it could be some complicated function of the spot values over time, with an average, as a typical example.

We consider the following payoff types:

a. **Vanilla Payoff:** Recalling Equations (4.1) and (4.3), we could write:

$$\max\{0, \eta(S_T - X)\} = (\eta(S_T - X))^+, \qquad (16.1)$$

where $\eta = 1$ for call options and $\eta = -1$ for put options, S_T denotes the spot value at time, T, and X is the exercise value.

b. **Binary Payoff:** A fixed payoff that only depends on whether the option is in-the-money, or not, that is,

$$Call = \begin{cases} 0, & \text{if } S_T \leq X \\ 1, & \text{otherwise.} \end{cases}$$

Binary options find application for rebates attached to barrier options, for example. Binary (or digital) options are difficult to hedge. Taleb [100] cautions:

> At the heart of most exotic structure and every bet resides a binary.

De Weert [41] notes:

> A call spread is not only used to price a digital, but from the perspective of the trader, it is also the product that he actually trades. In other words, when a trader sells a digital he books a call spread in his risk management system instead of the exact terms of a digital.

We could also have a binary asset payoff, that is,

$$Call = \begin{cases} 0, & \text{if } S_T \leq X \\ S_T, & \text{otherwise.} \end{cases}$$

c. **Gap Payoff:** The option only has a payoff if it finishes a certain depth into the money. Hence,

$$Call = \begin{cases} 0, & \text{if } S_T \leq (K \geq X) \\ S_T - X, & \text{otherwise.} \end{cases}$$

d. **Power Payoff:** The payoff is given by some function of the intrinsic value of the option, that is,

$$\max\{0, S_T - X\}^\alpha,$$

where $\alpha > 0$; alternatively,

$$\max\{0, S_T^\alpha - X\}.$$

e. **Self-quanto Payoff:** So-called self-quanto payoffs would be given by:

$$S_T \max\{0, S_T - X\}.$$

Note, all the payoff formulae above could be modified to reflect the payoff profile of a put option, I have merely used a call-type payoff for convenience.

16.3 EXOTIC VARIETIES: THINKING ABOUT PRODUCTS

Based on our previous analysis, we know the value of a Vanilla European option is a function of:

1. **Spot:** Current observed spot rate, S_t.
2. **Strike:** Predetermined strike rate, X.
3. **Tenor:** Tenor (lifetime) of the option, $T - t$.
4. **Carry:** Risk-free rate(s) of return, r and q.
5. **Volatility:** Volatility of the underlying, σ.

Purely as an aid to understanding and thinking of new products and different varieties of products and options, we classify exotic options based on the 'primary' change to the inputs of variables in the vanilla European option. In the next sections, we detail our thinking.

16.4 SPOT-RELATED CHANGES

This class of exotic option is obtained by attaching a new meaning to the way the spot rate is calculated.

1. **Average Rate Options:** The spot rate of the option is determined by looking at some average of underlying rates obtained in the market over the lifetime of the option. Typically, the effect would be to reduce volatility. The payoff of the option would be given by:

$$\max\{0, \frac{1}{N+1} \sum_{i=0}^{N} S_{t_i} - X\},$$

   for a discretely sampled call. See Equation (16.2) for the calculation pertaining to the continuous average.

2. **Lookforward Option:** The spot rate used to settle the option is determined by considering the maximum difference between the spot rate and the strike rate during the life of the option. This option could also be referred to as a fixed-strike lookback option.

16.5 STRIKE–RELATED CHANGES

This class of exotic option is obtained by attaching a new meaning to the strike of the option. We actually give the user the flexibility of not choosing the strike by 'allowing' the market to determine the choice.

1. **Floating Strike Lookback Options:** The strike of the option is determined at maturity by determining the min/max of the underlying over the life of the option. The payoff would be given as

$$\max\{0, S_T - \min\{S_{T_0}, S_{T_1}, \ldots, S_{T_n}\}\},$$

   for a lookback call. Here the S_{T_i} denotes discrete fixings of the underlying spot price, typically observed at the close of business or at a specified time. The lookback could obviously be observed 'continuously' (i.e., by tick-moves) as well, but with increased administration and system cost.

As an extreme example, we could have a lookback straddle with payoff depending on the difference between the maximum and minimum of the spot levels during the life of the option. We could also have a so-called *swing option* with call payoff:

$$\max\{0, (max_{t_1} - min_{t_2}) - X\},$$

where max_{t_1} and min_{t_2} denote the maximum/minimum spot levels over the lifetime of the option, respectively.

2. **Average Strike Options:** The strike of the option is determined by looking at a set average of rates obtained in the market over the lifetime of the option. The payoff of a continuously sampled call option (on the underlying spot), for example, would be given by:

$$\max\{0, S_T - \frac{1}{T}\int_0^T S(\tau)d\tau\}. \qquad (16.2)$$

Practically, it would be onerous to calculate the integral in Equation (16.2). These options are, therefore, normally observed at discrete time periods in practice; for example, at the close of trading. The same comments apply as with the lookback option.

3. **Shouter Options:** The strike of the option is determined during the life of the option – when the user believes the market has bottomed, say, for call options.

   These options are often encountered in structured products. The shouting feature might be embedded either algorithmically or contractually.

4. **Forward Starting Options:** The user receives the option only at some predetermined time in the future, say T_1. The strike of the option will be set at the ruling spot, S_{T_1}, and yield the following payoff:

$$\max\{0, S_{T_2} - \alpha S_{T_1}\},$$

for $\alpha > 0$ and $T_1 < T_2$. Recall Equation (10.14), which is valid for $\alpha = 1$.

   Forward starting options appear deceptively easy to price and hedge. However, as de Weert [41] notes:

For a long period several banks have been pricing these forward starting options using the skew of the specific maturity rather than the skew associated with the maturity equal to the term of the forward starting option, i.e., the maturity minus the forward start date. This way of pricing has resulted in large losses for some banks.

Forward starting options find great application in so-called ratchet or cliquet options. Cliquet options are one of the most popular members of the guaranteed fund family, for example, capital-protected investments.

In its simplest form, the product pays off the yearly increases of an underlying index, and ignores any decreases. In the case where we have the payoff taking place on maturity, we can write:

$$\sum_{t=1}^{N} N_t \max\left[0, \frac{S_t - S_{t-1}}{S_{t-1}}\right].$$

Here S_t denotes the closing level of the index per period and N_t denotes the nominal amount exposed per period. From this formula, we can see that in this form the cliquet option is effectively a series of forward starting options. A very popular extension to the cliquet structure is to place a cap on the maximum payoff per period, for example,

$$\sum_{t=1}^{N} \max\left[0, \min\{\frac{S_t - S_{t-1}}{S_{t-1}}, \text{CAP}\}\right],$$

where CAP denotes the global cap-level imposed on the structure.

16.6 CHANGES AFFECTING THE OPTION MATURITY

This class of exotic option provides for an American-type characteristic in the option depending on the behaviour of the spot. The end user can typically lose or gain the option.

1. **Compound Options:** Options on options. The user fixes a price where an underlying option can be bought/sold. Here the payoff is:

$$\max\{0, C(S_t, T-t, X) - X_c\},$$

for a call (strike X_c) on a call (strike X), as an example. Taleb [100] cautions:

> Compound options are extremely sensitive to higher derivatives with respect to spot, particularly the fourth moment of the distribution, which traders call volatility of volatility.

2. **Barrier Options:** Knock-in/-out varieties provide for a level of flexibility where the user could gain/lose the option depending on whether some barrier level is breached. Rebates are frequently attached to these options, resulting in a binary payoff when the option knocks out or does not knock in.

   An up-and-in call option would, therefore, have payoff ($S_0 < H$, where H is the barrier level):

   $$Call = \begin{cases} \max\{0, S_T - X\}, & \text{if for some } \tau \leq T, S_\tau \geq H \\ 0, & \text{otherwise.} \end{cases}$$

   An up-and-out call would have payoff:

   $$Call = \begin{cases} \max\{0, S_T - X\}, & \text{if for all } \tau \leq T, S_\tau < H \\ 0, & \text{if for some } \tau \leq T, S_\tau \geq H. \end{cases}$$

   In Table 16.1, we compare prices for European call options to those of European down-and-in call as well as down-and-out call prices, respectively. Note that the down-and-in and down-and-out prices sum to the respective call prices. The barrier level is denoted by H.

   a. **Trading Example:** Barrier options have numerous applications in practice. Consider the example of an asset manager who wants to hedge their portfolio based on his short-term view that the market might pull back even though who? they market? might be longer-term bullish. Purchasing an outright put would be expensive, but purchasing an up-and-out put would be considerably cheaper.

   b. **Structures:** Barrier options are also used in so-called ladder structures. These structures serve to protect gains collected in the market based on certain levels being reached.

TABLE 16.1 Comparing a European Call with Down-and-In As Well As Down-and-Out Call Option Prices: $T = 1, \sigma = 20\%, r = d = 0\%, X = 90, H = 80$

Spot	Down-and-in	Down-and-out	European Call
100.00	0.34	13.25	13.59
97.50	0.46	11.35	11.81
95.00	0.61	9.53	10.13
92.50	0.80	7.78	8.59
90.00	1.05	6.12	7.17
87.50	1.37	4.52	5.89
85.00	1.78	2.97	4.75
82.50	2.28	1.48	3.76
80.00	2.91	0.00	2.91
77.50	2.19	0.00	2.19
75.00	1.61	0.00	1.61
72.50	1.15	0.00	1.15
70.00	0.79	0.00	0.79

c. **Monitoring:** It is generally very important to understand how barrier options are monitored against their knock-in/-out features. Parisian options are an important variation of barrier options, where the barrier is only breached on an average basis or dependent on the time the spot spent above/below a barrier level.

De Weert [41] notes:

> Barrier options are very popular amongst retail investor as the barrier feature provides the investor with additional protection or leverage. From a risk management perspective, barrier options are interesting because the risks are discontinuous around the barrier and therefore the Greeks become less predictable and very often even change sign around the barrier.

3. **Chooser Options:** Options for the undecided, which give us the flexibility of choosing, at some future time, whether a call or a put was bought.

4. **Contingent Options:** The option is only paid for if it is in-the-money at expiry.

16.7 MULTI-ASSET OPTIONS

Multi-asset options effectively involve changes with regard to the underlying. When we consider so-called multi-asset options, correlation will become a central theme; it, effectively, influences the volatility of the total option portfolio. These options range from basket options to rainbow options. In principle, though, we are influencing the effective underlying assets, and view it as a change of the effective underlying.

1. **Basket Options:** The volatility of a basket of assets is typically reduced because of the correlation between these assets. Basket options provide cheaper insurance than buying individual options on individual assets.

   The payoff of a basket call option (strike X) would typically be

   $$\max\{0, \sum_{i=1}^{N} w_i S_i - X\},$$

   if we are given a basket of assets $S_1, \ldots, S_N$ with respective weights $w_1, \ldots, w_N$ subject to $\sum_{i=1}^{N} w_i = 1$.

2. **Rainbow Options:** Provide for the flexibility of choosing the best performing of two or more risky assets over time.

   a. Typical rainbow options on assets $(S_1, \ldots, S_N)$ could include

   $$\max\{S_1, \ldots, S_N\},$$

   a so-called N-colour better-of option.

   b. Another example could be

   $$\min\{S_1, S_2\},$$

   a so-called two-colour worst-of option.

   c. A well-known example would be an out-performance option on two assets

   $$\max\{S_2 - S_1, 0\},$$

easily extended to
$$\max\{w_2 S_2 - w_1 S_1, 0\},$$
for weights w_1, w_2.

3. **Linked Domestic and Foreign Exposures:** If we consider the global equity markets, for example, we are able to link foreign stock and currency exposure in some interesting ways.

    Abusing previous notation, we denote the currency at time, T, by X_T, that is, the forex translation rate. The foreign equity will similarly be denoted by S'_T. Let us now consider four different payoff scenarios. These reflect different views on risks varying between equity and currency exposure.

    a. **Foreign Equity Call Struck in Foreign Currency:** An investor wishes to participate in gains in a foreign equity; hence, she buys a call, but has no currency translation concerns. In this case, the following payoff is desirous:
    $$X_T (S'_T - K')^+,$$
    where K' denotes a strike in foreign currency.

    b. **Foreign Equity Call, Struck in Domestic Currency:** For an investor who is concerned about the accounting currency value of a foreign investment, the following option payoff is desirous:
    $$(S'_T X_T - K)^+,$$
    where K denotes a strike in domestic currency.

    c. **Fixed Equity Call, Struck in Domestic Currency:** This combination, frequently referred to as a *Quanto*, will have payoff:
    $$\bar{X}(S'_T - K')^+ = (S'_T \bar{X} - K)^+,$$
    where $\bar{X}$ denotes a fixed translation value, that is, foreign currency translation risk has been removed.

    d. **Equity-linked Foreign Exchange Call:** In this scenario, an investor holds foreign equity exposure with no concern, but wishes to protect the currency translation risk, that is,
    $$S'_T (X_T - K)^+.$$

16.8 REMARKS AND FURTHER READING

We have shown the tip of the iceberg. Various other exotics exist and are traded world-wide. Different exotic option varieties are also used (and customised) in different markets.

1. **Hedging:** It should be evident that these options all share some path-dependency, which typically creates more hedging complexity than hedging vanilla options. Taleb [100] is an essential read. Also read [14, 41, 92].

2. **Popularity:** In some respects, we have also gone full circle as some of the exotics described above have become *vanilla* as the markets have matured and some of these options have gained in popularity. Barrier options, for example, are traded in almost all areas of modern derivative financial markets.

3. **Extensions:** Using the methodology outlined above, we could easily allow for combinations, so the following examples might be possible:

    a. Average-rate digital knockout basket options with rebates.

    b. Range-bound cliquet options on an index based on a roller-coaster nominal with a quanto into the best performing of a basket of currencies chosen at inception.

    c. Bermudan lookback knock-in power options.

4. **Financial Engineering:** Exotic options are true signs of the innovative power of modern finance [25]. These options present challenges on many levels of an organisation, ranging from risk management to administration to pricing and system design. They are, however, part of modern finance and in practical use. Most structured products traded world-wide would not be possible without some exotic feature being enabled, for example.

    The essence of Financial Engineering is, therefore, shown here. We identify financial problems and solutions – in order for us not to have to absorb the risk of proposed solutions we need to ensure that we have priced solutions as accurately as possible and, tied to that, we need to have a viable hedging programme – these two ideas frequently go hand-in-hand.

    Please remember, we aim to solve financial problems. Exotic derivatives are examples of such solutions. It is extremely important, though, to remember that we want to determine accurate pricing and that this is accompanied by a robust replication (or hedging) programme. Without a clear understanding of our hedging methodology, the pricing is rendered meaningless!

5. **Embedded Derivatives:** It is also very important to understand that many commercial contracts, in practice, can be decomposed into the analysis of contingent claims, that is, many contracts contain so-called real-options or some exotic-type payoff embedded in them. See [51, 72], for example. These contracts need to be appropriately valued and frequently require some hedging technology. Examples include longer-term contracts where we sell a product (such as oil at a fixed price or contingent at some level being reached or contracts that give a conditional guarantee based on an underlying index subject to mortality, or insurance type risks).

6. **Pricing Concerns:** Some caution is in order. Exotic options are frequently priced in the Black–Scholes framework and are, therefore, subject to the assumptions used within this framework. Some assumptions, for example, the availability of liquidity, become crucial for exotic options where the Gamma of the option could become exceedingly large or change sign rapidly. While it is theoretically possible to price any options we can dream of, the hedging of such options might turn into a nightmare! Read Taleb [100].

   We have also paid no attention to some crucial market calibration issues such as the volatility surface or correlation aspects. See, [2, 40, 41, 50].

   Note that we frequently assume the Black–Scholes–Merton framework [9, 79]; however, in practice, numerous techniques are used for pricing, the binomial option pricing framework [34, 54] being one popular example. We also discussed the Monte Carlo method in Section 10.8. This method is of significant use in practical applications. See, for example, [15, 17, 29, 55, 103].

   Practically, we also need to extend our thinking to other settings such as jumps (see, [21, 81], for example) and stochastic volatility ([48, 50], for example), to extend the basic Black-Scholes-Merton paradigm. Calibration issues are of fundamental importance, see, for example, [14, 50, 52, 97].

7. **Monitoring:** Most contracts, in practice, are formulated in terms of discrete monitoring. See, for example, [18]. Lookback and Asian-type options could be monitored at the close of business, for example. In FX, we could have more regularly monitored barriers; it is extremely important to understand what constitutes breaching a barrier, that is, do we need to observe trading at or below a down-and-out barrier, for example, or merely see an offer below the barrier?

CHAPTER 17

Model Validation Process

It is by no means obvious which model is correct, nor even exactly what correct means.

– EMANUEL DERMAN

Euclid taught me that without assumptions there is no proof. Therefore, in any argument, examine the assumptions.

– ERIC TEMPLE BELL (1883–1960)

The Modelers' Hippocratic Oath

I will remember that I didn't make the world, and it doesn't satisfy my equations. Though I will use models boldly to estimate value, I will not be overly impressed by mathematics. I will never sacrifice reality for elegance without explaining why I have done so. Nor will I give the people who use my model false comfort about its accuracy. Instead, I will make explicit its assumptions and oversights. I understand that my work may have enormous effects on society and the economy, many of them beyond my comprehension.

– EMANUEL DERMAN, PAUL WILMOTT (JANUARY 7, 2009)

Model Validation Process ▪ 167

17.1 MOTIVATION

Based on the earlier chapters, we now have a sense of building a derivative model based on some established principles. We rarely, however, build models just 'for the sake of it'. We have users and we need to ensure we deliver a reliable, useful, and correctly working product. We refer to the process of implementing, testing, verifying, and conferring conclusions as model validation. Our examples are specific to derivative securities but the principles apply to any modelling endeavour.

Let us start by showing a relevant, practical example, detailing how model implementation could create problems.

17.2 INTRODUCTORY EXAMPLE

Consider a call option on a stock that pays no dividends; we recall the fair Black–Scholes option price, that is, the solution to Equation (12.4) is given by:

$$C = SN(h^+) - Xe^{-rT}N(h^-), \tag{17.1}$$

where $h^\pm$ is given by:

$$h^\pm = \left[\ln(\frac{Se^{rT}}{X}) \pm \frac{\sigma^2}{2}T\right] / \left(\sigma\sqrt{T}\right), \tag{17.2}$$

subject to the known (and agreed) parameters $S, X, r, \sigma,$ and T. Furthermore, the cumulative Normal distribution function is defined as

$$N(x) = \frac{1}{\sqrt{2\pi}} \int_{-\infty}^{x} e^{-s^2/2} ds. \tag{17.3}$$

We note, in passing, that there is generally no closed-form expression for $N(\cdot)$. We would typically read values off a table or use some numerical approximation to obtain values. (Consider, for example, the use of NORMSDIST() in EXCEL that is based on a polynomial approximation method.)

Let us use an approximation from Abramowitz and Stegun [1]:

$$N(x) \approx 1 - \frac{1}{2}\left(\sum_{i=0}^{4} c_i x^i\right)^{-4} + \epsilon(x), \tag{17.4}$$

where $|\epsilon(x)| < 2.5 \times 10^{-4}$ and

$$c_0 = 1.000000,$$
$$c_1 = 0.196854,$$
$$c_2 = 0.115194,$$
$$c_3 = 0.000344,$$
$$c_4 = 0.019527.$$

Seems innocent, right? Let us apply this approximation to the Black–Scholes model, given by (17.1) and (17.2), to calculate the value of an out-of-the-money call option. For illustrative purposes, we assume the following values $S_0 = 100$, $X = 125$, and $T = 0.5$ with $r = 10\%$ and $\sigma = 25\%$.

The results displayed in Table 17.1 are not great; the approximation is barely accurate to one decimal, and worryingly, creates arbitrage opportunities. Call option prices cannot be negative? Recall the discussion in Chapter 6, Equation (6.1).

The lesson here is that implementation is important and we need to understand all aspects of a model on a detailed basis to ensure we do not violate basic principles.

TABLE 17.1   Call Option Approximation vs Correct Analytical Values for a Range of Spot Values

Spot	Approximate Value	Analytical Value
50	−0.000235978082	0.000001232216
55	−0.000743535641	0.000019320514
60	−0.001963889838	0.000188421239
65	−0.003910870701	0.001259436022
70	−0.003789784713	0.006205548750
75	0.009830830995	0.023833583088
80	0.063740828322	0.074533215402
85	0.199631260744	0.196476748143
90	0.466194176945	0.448952275547
95	0.921771705749	0.909729350873
100	1.653183195635	1.665749773518
105	2.776830526650	2.799551222742
110	4.387272089228	4.375712876591
115	6.460031017273	6.431340242061
120	8.929662550574	8.972697216543

Let us use another approximation from Abramowitz and Stegun [1]:

$$N(x) \approx 1 - \frac{1}{2}\left(\sum_{i=0}^{6} d_i x^i\right)^{-16} + \epsilon(x), \qquad (17.5)$$

where $|\epsilon(x)| < 1.5 \times 10^{-7}$ and

$$d_0 = 1.0000000000,$$
$$d_1 = 0.0498673470,$$
$$d_2 = 0.0211410061,$$
$$d_3 = 0.0032776263,$$
$$d_4 = 0.0000380036,$$
$$d_5 = 0.0000488906,$$
$$d_6 = 0.0000053830.$$

The results in Table 17.2 are much better, but when we extend the range of option values lower, we find that option values can still become negative. Again, the lesson here is that implementation matters! Implementation of models can create unwanted risks.

Inspired and motivated by this example, it is clear that we need a framework to assess the soundness of models, that is, their implementation

TABLE 17.2 Call Option Approximation vs Analytical Values for a Range of Spot Values

Spot	Approximate Value	Analytical Value
30	0.000000000000	0.000000000000
35	−0.000000000031	0.000000000004
40	−0.000000001304	0.000000000671
45	−0.000000002414	0.000000042802
50	0.000000759384	0.000001232216
55	0.000016847234	0.000019320514
60	0.000181696225	0.000188421239
65	0.001250728261	0.001259436022
70	0.006204145066	0.006205548750
75	0.023842812042	0.023833583088
80	0.074539218077	0.074533215402
85	0.196468097526	0.196476748143
90	0.448943629209	0.448952275547
95	0.909738570615	0.909729350873
100	1.665759911925	1.665749773518

and applicability, which is more completely described as validation. In the words of William P. Thurston,

> Mathematics is not about numbers, equations, computations, or algorithms: it is about understanding.

In a 2002 report [101], the NIST estimates that software errors cost the US economy $59.5 billion annually. While validation entails much more than just testing (or verification), it illustrates the quantum of importance for this whole subject matter.

17.3 SCOPE

Our aim is to validate a model. As the example in Section 17.2 shows implementation matters! But the total validation of a model requires an understanding of what is required to be validated. As Richard Feynman noted:

> The first principle is that you must not fool yourself, and you are the easiest person to fool.

It is therefore extremely important to define the scope of our model validation exercise. This might be different depending on the company, users, and, even, regulatory approaches.

The Board of Governors [13] has this useful definition of a model (specifically quantitative models) in their SR 11-7 guidance note:

> ..., the term model refers to a quantitative method, system, or approach that applies statistical, economic, financial, or mathematical theories, techniques, and assumptions to process input data into quantitative estimates. Models meeting this definition might be used for analyzing business strategies, informing business decisions, identifying and measuring risks, valuing exposures, instruments or positions, conducting stress testing, assessing adequacy of capital, managing client assets, measuring compliance with internal limits, maintaining the formal control apparatus of the bank, or meeting financial or regulatory reporting requirements and issuing public disclosures. The definition of model also covers quantitative approaches whose inputs are partially or wholly qualitative or based on expert judgment, provided that the output is quantitative in nature.

The Board of Governors [13] also states:

> A model consists of three components: an information input component, which delivers assumptions and data to the model; a processing component, which transforms inputs into estimates; and a reporting component, which translates the estimates into useful business information.

Let us now discuss these aspects in more detail.

1. **Model Inputs:** We can observe some model inputs directly in the market (e.g., equity, and bond prices, swap rates, and traded implied volatilities), whereas some model inputs (e.g., forward volatilities and correlations) may not be directly observable, and are inferred from other observable quantities. (These might even be calculated using an algorithm based on some assumptions of the state of the world.)

    Importantly, inputs that are not directly observable potentially give rise to further model uncertainties. In a typical bank trading setting, we would, for example, consider capital buffers or provisions for these type of inputs.

2. **The Analytics:** The analytics (which typically incorporates the model paradigm and implementation) transform the inputs into a value for the product.

    a. Some analytics are standard and straightforward (e.g., calculating the PVBP of a bond).

    b. The analytics of some models can be more complex – such models are typically based on a body of theory that makes some assumptions about the state of the world such as the distributions of the underlying market factors.

    c. Since any model is only an estimate of the real world at best, it is important to understand the model limitations with regard to the appropriateness of its assumptions.

    In practice, we make business decisions to participate in certain products and markets; we need to ensure our understanding and ensure we have the necessary risk appetite for an activity. Therefore, we will need to ensure that the necessary financial provisions are taken to enable these business activities.

d. Additionally, the implementation of the assumptions may involve mathematical techniques such as Monte Carlo simulations, which have their own limitations and may have a significant impact on the results.

3. **Model Outputs:** The outputs of the model are typically used for profit and loss reporting, exposure and hedge management, books of accounting and records for the company, and risk management monitoring and capital requirements.

17.4 MODEL VALIDATION

Model validation requires a full business process. In the spirit of the inspirational words of Beverly Sills,

> There are no shortcuts to any place worth going.

A derivatives model would typically include the following aspects:

1. **Client Promise:** A complete understanding of the client promise. This should include:

    a. A full description of legal aspects pertaining to the derivative's payoff.

    b. It is important to understand if settlement of any contingent claim is physical or in cash.

    c. A full description of how path-dependency is handled.

    d. What is the process of handling early-exercise and its implications?

    e. What happens if there is a default event?

    f. Any contractual margining or collateralisation.

    g. An understanding of any early unwinding or termination of the product.

2. **Assumptions:** An understanding of the general assumptions arising from a chosen paradigm to create a mathematical model for pricing and hedging of a contingent claim.

3. **Implementation:** A complete implementation of the mathematical model to produce prices and hedging ratios. This includes all computational aspects, which frequently entails making some approximations.

4. **Inputs:** The correct interpretation of inputs and calibration to reproduce known market prices from a model.

5. **Outputs:** The correct interpretation of all outputs, prices, and Greeks.

6. **Risk Management:** We use models to calculate (or estimate) the value of derivative securities. The hedging strategy, also referred to as the replicaton strategy, is equally important. We, therefore, seek an understanding of the replication strategy (and risk management) implied (or required) by the model.

The Board of Governors, guidance on model validation [13] is as follows:

> Model validation is the set of processes and activities intended to verify that models are performing as expected, in line with their design objectives and business uses. Effective validation helps ensure that models are sound. It also identifies potential limitations and assumptions, and assesses their possible impact.

Therefore, model validation generally consists of the following:

1. **Review the Theoretical Appropriateness of the Model:** Once we have chosen our paradigm, we move to models for specific instruments. A model will attempt to give the fair value of a derivative asset/liability. It should also supply information on how the derivative should be replicated.

2. **Assess the Reasonability of the Proposed Modelling Approach Applied to the Product:** Simple derivatives, such as stock options, for example, have models that are well accepted by the trading community and these models are used to assist in the quoting conventions of these derivative products. This does not necessarily mean these models are correct.

3. **Confirm that the Model is Correctly Implemented:**
    a. Based on the specific paradigm chosen, we have closed-form solutions to certain derivatives. Closed-form solutions are attractive

and provide a sense of completeness. The formulae are nevertheless non-trivial and one should test the formulae exhaustively for accuracy. Furthermore, closed-form solutions should be verified from first principles as there are frequently mistakes in textbooks. One should also understand the numerical approximations used to implement closed-form solutions. Recall the cumulative normal distribution example in Section 17.2. Frequently, approximations of this distribution have limited accuracy.

b. Where numerical solutions are relied upon to solve problems, a careful study should be made to ensure convergence of the numerical technique. There are commercial vendors, for example, who use binomial techniques to price derivatives but hard-code the number of steps used in the algorithm.

c. To ensure adherence to basic financial principles, numerical methods should be carefully tested. Commercial codes, which implement binomial option pricing methods that violate put–call parity, for example, exist (recall, Section 9.3). One should also ensure that the numerical method is consistent with the original chosen paradigm. Numerical approximations could lead to paradigm shifts.

d. We also need to assess the appropriateness of the calibration of our model to real-world products and related instruments.

4. **Identify Any Model Limitations and Uncertainties, Including Model Parameters, that Impact Valuation:** It is important to ensure a broad range of tests on a specific model. Typical tests should 'stress-test' models. Boundary conditions should be tested. As an example, we should observe the model for 'pathological' values of the underlying. As examples:

    a. What happens when the underlying tends to zero or infinity?

    b. What happens at, or near, a barrier?

    c. What happens on a reset date?

    d. What happens at expiry?

Most exotic derivatives are highly nonlinear, potentially even discontinuous, with respect to movements in the underlying. It is important to understand and verify the response of a model to changes in the underlying parameters.

5. **Identify the Primary Risks and Quality of the Risk Parameters Calculated:** The following advice from Taleb [100] is relevant in this context:

> Traders and managers need to make a clear distinction between the contractually path-dependent positions (i.e., barriers) and the securities that are path dependent owing to dynamic hedging (vanilla options).

In short, all options have some form of path-dependency as a result of the way we replicate and hedge them. It is, therefore, important to simulate models through different life cycles. We should test the model and its hedge parameters from the inception of a deal until the expiration of the deal. It is important to gauge whether the hedging technique implied by a model does indeed lead to a replicated derivative in some sense.

Hedging could also involve some form of basis risk. As an example, we could have a contingent claim on a specific stock but hedge it using a general index. As another example, we could have a named credit-default swap (CDS), which we hedge using a general graded index CDS.

6. **Recommend Appropriate Provisions to Account for the Model Limitations and Uncertainties:** Once we have identified the limitations of a model, there is a decision faced by the business whether we continue with the specific line of business, change the model, or build provisions to account for the uncertainty.

17.5 VALIDATION METHODOLOGY AND PROCESS

We now turn to the validation process. In the words of software developer, Tim Peters:

> The proof of the pudding is in the eating, and the proof of a model is in its validation.

Let us assume we have agreed to validate a model. Broadly speaking, three possible cases exist and the validation process for each builds on the previous approaches.

First, consider the problem of validating a model for which a closed-form solution exists (E.g. validate the implementation of a Black–Scholes option pricing model). Second, consider validation of a numerically implemented model within a known paradigm (E.g. validate a finite-difference approximation to a three-factor stochastic volatility implementation). Third, consider the validation of a 'Black Box' (E.g. a software vendor provides a model for valuation of convertible bonds). We shall consider the validation process for these problems sequentially. The approach should be seen as incremental and inclusive.

In all cases, the validation (which includes verification) approach taken should include an independent model built by the person performing the testing. The independence of this model is clearly crucial.

Ideally one should decide on error tolerance before starting the evaluation. The tolerance should be chosen so as to reflect the dimensionality and path dependency of the problem, while bearing the analytic tractability of the problem in mind. On the one hand, one should expect error tolerances within machine accuracy for analytic solutions but on the other hand much bigger errors for a Monte Carlo simulation, for example.

Now, consider the different cases described above. Sherlock Holmes always has applicable insights:

> But this case, my dear fellow, is, of course, a patent and even egregious example of observation and deduction.

CASE A: Firstly, consider a model for which closed-form analytical solutions are known. We need to answer the following questions to satisfy ourselves of its implementation.

1. How has the analytical solution been derived? Can we replicate the formulae?

2. Do we fully understand the inputs to our models? (This might seem trivial but what are the input conventions, that is, 20% or 0.2?)

3. Do we need to make approximations to evaluate analytical expressions? If we do, how accurate are those approximations?

4. Are we able to reproduce the same results as per the model's solution for a wide range of inputs?

5. How do we test 'pathological' cases? Examples are:

   a. The value of an option at expiry.

   b. Behaviour of a model for very small/large values of the underlying.

   c. Compliance with boundary conditions, for example, barrier options.

   d. Understanding the path-dependent characteristics of an option. As an example, what happens at the resets of an Asian option?

   e. Are there areas where the solution does not exist?

   f. How do we cater for discrete dividends? Have we tested what happens when the present value of dividends exceed the current spot value?

6. Of equal importance to accurate calculation of prices would be the accurate calculation of the Greeks. These are frequently cumbersome to evaluate analytically; hence, approximations are often used in their evaluation. One should ensure that the same process is followed to ensure the accuracy of the Greeks as with the model prices.

7. In general, for closed-form solutions, we should be able to replicate the results of a model implementation to within machine accuracy of the correct analytical values. If answers between the model and our test model differ, we should investigate the difference thoroughly and be able to explain the differences analytically.

CASE B: Consider the validation of a model that has been implemented using a numerical method. Clearly, the principles highlighted for CASE A still hold and need to be adhered to; however, we also need to consider the following questions:

1. Within an existing paradigm, how has the problem been discretised numerically?

2. Determine the behaviour of the model with respect to discretisation parameters. Does the implementation converge to a single number as the discretisation parameter becomes 'arbitrarily' small?

3. Is the model implementation stable with respect to small movements in the underlying?

4. Does the numerical accuracy suffer from an increase, or decrease, in time?

5. Describe the solution for a wide range of inputs. Does the solution appear jagged or smooth? Explain the Greeks and examine their behaviour.

6. Examine errors between the model and a test model. Are the relative errors stable or widely fluctuating?

7. Are the differences between the models reconcilable, that is, can we attribute differences to numerical approximations, for example?

8. Testing can become problematic from a time perspective in cases where the numerical solution relies on simulation or other time-consuming computational techniques. In this situation, one should plan the verification testing more carefully to ensure a broad-based verification with a smaller sample of tests being conducted.

CASE C: The last case we consider presents the most headaches. In this case, we are validating a 'Black Box'.

It is nearly impossible to validate a 'Black Box' method in general. Frequently, we might not even understand the 'Black Box' paradigm. The number of variables to test against makes reasonable testing very involved. We need to apply our principles outlined above, though.

The first step pertains to our understanding of the problem and choice of a method to test the 'Black Box' against. It should be clear that in this verification process we could only give a qualified verification at best. The reason being that we test against another model, which becomes difficult to prove superior to the Box. We need to ensure that our test model adheres to the principles contained above.

17.6 CONCLUSION AND FURTHER READING

I read a book recently that had too many characters and no plot. When I brought it back to the library they said, "Why did you steal the phone book?"

– DOROTHY FRASER

Model validation in Financial Engineering is no different from the validation process followed in Physics or Engineering. We use mathematical model information and decision-making purposes and attach significant weight and credibility to their predictions and insights. Mistakes could lead to catastrophic consequences. See, for example, [37, 38, 101, 103].

It should be clear from the above that validation is as important as building a model! The model validator needs to be able to build an alternative model to compare results against and be able to understand the intricacies of a model. This process shows how important clear documentation of a model becomes. The scientific approach should ensure full replicability. In our context, the validation of derivative prices is by no means a trivial process. We need to ensure a proper understanding of all assumptions made in the derivation of a price and everything associated with that price.

The process we have described is not restricted to derivatives model and could be applied on any models. It is important to understand the current thinking in regulatory and accounting circles along with increased corporate governance. See, e.g., [13]. This will force us to ensure a strict process is followed to validate derivative prices.

The governance of models and the documentation thereof is fundamentally important. In Chapter 18, we discuss the risk management process in some detail. Model validation, generally, would form part a of a risk management process.

CHAPTER **18**

Risk

Risk means more things can happen than will.

– ELROY DIMSON

The expected never happens; it is the unexpected always.

– JOHN MAYNARD KEYNES (1883–1946)

The revolutionary idea that defines the boundary between modern times and the past is the mastery of risk: the notion that the future is more than a whim of the gods and that men and women are not passive before nature.

– PETER L. BERNSTEIN (1919–2009)

The trouble with risk management is that if nothing happens, then people think you're a fool, and if something does happen, people think you're a fool.

– EMANUEL DERMAN

A man who carries a cat by the tail learns something he can learn in no other way.

– MARK TWAIN (1835–1910)

18.1 INTRODUCTION

Over the last four decades, multibillion dollar losses in the financial markets have served to focus the attention on adequate management of market risks. In many cases, these losses have arisen from new investment-related product innovation, or a mere misunderstanding of the risks inherent in a specific (new) business endeavour, involving the financial markets.

Let us consider a couple of financial disasters in recent years. In all of these cases, we see some derivatives involvement in common.

1. **Metallgesellschaft (1993):** Metallgesellschaft AG, a German industrial conglomerate, suffered significant losses due to its subsidiary's disastrous oil trading activities. The subsidiary, MG Refining and Marketing (MGRM), entered into long-term oil supply contracts with fixed prices, expecting to profit from the contango market structure. However, when oil prices unexpectedly fell, MGRM faced massive margin calls that it could not meet. Metallgesellschaft had to step in to cover the losses of more than USD 1 billion, which led to to a major financial crisis for the company and its eventual restructuring. See, [44, 87], for an overview and discussion.

2. **Barings Bank (1995):** Nick Leeson, a derivatives trader at Barings Bank in Singapore, engaged in unauthorised trading and concealed losses through an error account. Leeson's risky bets on Nikkei futures and options eventually led to the bank's collapse, making it one of the most infamous financial scandals in history. Approximately USD 1.3 billion loss at the time.

3. **Sumitomo Corporation (1996):** Yasuo Hamanaka, a trader at Sumitomo Corporation, manipulated the copper market by amassing large positions in copper futures and options over a period of 10 years. Hamanaka's unauthorised trading led to substantial losses, approximately USD 2.6 billion, for Sumitomo when the scheme was uncovered.

4. **Long-Term Capital Management (LTCM) (1997):** LTCM, a hedge fund led by Nobel laureates and Wall Street veterans, heavily invested in complex derivatives, particularly in fixed-income markets. However, their risk-management models failed amidst the Russian financial crisis and other market shocks, which led to massive losses and required a bailout orchestrated by the FED. LTCM's losses amounted

to roughly USD 4.6 billion. See, [62], for an instructive discussion on the event.

5. **Enron (2001):** Enron employed off-balance-sheet special-purpose entities (SPEs) to hide debt and inflate profits. These entities were heavily involved in derivatives trading, particularly in energy markets. However, accounting irregularities and fraudulent activities were uncovered, leading to Enron's bankruptcy and the dissolution of Arthur Andersen, its auditing firm. The losses amassed amounted to roughly USD 74 billion. Consult Lowenstein [71] for an interesting account of the times.

6. **Amaranth Advisors (2006):** Amaranth Advisors, a hedge fund, made massive bets on natural gas futures and options, anticipating rising prices. However, unfavourable market conditions and poor risk management led to significant losses, roughly USD 6.6 billion, when natural gas prices declined sharply.

7. **Global Financial Crisis(GFC) (2007–2008):** The GFC has become a central theme of this century. The proliferation of complex derivatives, particularly credit default swaps, contributed to the subprime mortgage crisis. Financial institutions had amassed substantial exposures to mortgage-backed securities and related derivatives without fully understanding the risks involved. The collapse of Lehman Brothers and the subsequent global financial crisis highlighted the systemic risks posed by these instruments. See, Cassidy [24].

    The U.S. National Commission on the Causes of the Financial and Economic Crisis in the United States [102] provides an excellent reading source on this ground-shifting event. There remains no doubt that the 2007/2008 period was a defining period for risk management world-wide and will define the thinking for many years to come.

8. **JP Morgan Chase (2012):** JP Morgan's Chief Investment Office made large bets on credit derivatives, primarily credit default swaps, which were intended to hedge the bank's overall risk exposure. However, the trades, executed by a London-based trader nicknamed the 'London Whale,' were poorly monitored and grew to unsustainable levels, resulting in significant losses for the bank (approximately USD 6.2 billion was lost). See, [31], for the full detailed Congressional investigation.

9. **Options Clearing Corporation Margin Call (2020):** Amidst heightened market volatility triggered by the COVID-19 pandemic, the Options Clearing Corporation issued an unexpected margin call to its members, including major brokerages. The margin call, which had been intended to cover potential losses from increased trading activity and volatility, strained liquidity, and capital reserves at many firms, led to significant losses and operational challenges. Estimates of losses amount to several billion dollars.

10. **Archegos Capital Management (2021):** Archegos Capital Management, a family office, amassed highly leveraged positions in a handful of stocks through total return swaps and other derivatives. Margin calls were triggered when the value of these positions declined sharply, which forced brokers to liquidate Archegos' positions at steep losses amounting to several billion dollars. The losses resulted in significant financial fallout for both the fund and its counterparties.

11. **Silicon Valley Bank (2023):** Silicon Valley Bank Financial Group (SVBFG) was founded in 1983 and was headquartered in Santa Clara, California. Prior to its failure, SVBFG was a financial services company, and bank holding company with approximately $212 billion in total assets. The FOMC report on SVB [47] states:

    > The report shows that Silicon Valley Bank was a highly vulnerable firm in ways that both its board of directors and senior management did not fully appreciate. These vulnerabilities – foundational and widespread managerial weaknesses, a highly concentrated business model, and a reliance on uninsured deposits – left Silicon Valley Bank acutely exposed to the specific combination of rising interest rates and slowing activity in the technology sector that materialized in 2022 and early 2023.

    Furthermore, the report [47] details:

    > While low interest rates and more-frequent client funding events affected all financial institutions and their clients, SVBFG saw an outsized impact because of its concentration in venture capital and start-up clients, and SVBFG invested these deposits in long-dated securities. SVBFG's assets grew 271 percent from year-end 2018 to year-end

2021, compared to 29 percent for the banking industry. Asset growth slowed dramatically in 2022 as tech-sector activity slowed in a rising-interest-rate environment.

The bank, essentially, got caught between declining asset prices, resulting from interest rate increases induced by the central bank to tame inflation, and a cash-liquidity squeeze with disastrous consequences.

Voluminous books could be, and have been, written on the the list of losses above, and their related failures. It is important to understand that derivatives played a role in all these but fraud, and related behaviour, is not restricted to derivatives. What has been unique in most of these disasters has been the combination of a loss event and some derivatives with leverage, which has been toxic. Chance [25] provides prescient comments:

Since for every winner there is a loser, we would expect derivatives losses. What we would not expect is that an organization would have to announce that it had incurred losses, that these losses were inconsistent with expectations, were due to unauthorized trades, and/or that the organization would no longer be using derivatives.

18.2 RISK MANAGEMENT

Bouchaud and Potters [14] provide an epigrammatic description of risk management in the context of modern financial engineering:

Measuring and controlling risk is now a major concern in many modern human activities. The financial market, which act as highly sensitive (probably oversensitive) economical and political thermometers, are no exception. One of their rôles is actually to allow the different actors in the economic world to *trade* their risks, to which a price must therefore be given.

Risk management entails all aspects of a business; however, in this short chapter, we shall aim to present the essential ideas underlying a market risk-management programme. The reason for this choice is simply that market risk ties to the hedging and management of derivative securities, as well as the identification of related risks. Thus, we should not avoid the discussion.

Firstly, what is market risk? Market risk arises by virtue of our income statement or balance sheet changing in value as a result of changes in the values of assets and liabilities owing to changes in the value of underlying traded securities. Typically, these include (refer to Chapter 1):

1. Equities,
2. Foreign currency,
3. Interest rate(s),
4. Commodities,
5. Credit spreads, and
6. Changes in volatilities on the above instruments or associated correlations.

Let us now consider some pertinent examples of business endeavours that involve market risks:

1. **Investor:** Investors buy assets for long-term consumption and investment potential. Short-term market fluctuations create significant headaches that could question the investment case.
2. **Insurer:** Insurers frequently face valuation issues involving embedded promises in investment-related contracts. A simple promise to preserve capital could be seen as a derivative liability and give rise to significant income statement volatility as interest rates move around, for example.
3. **Market Maker:** A derivatives market maker, for example, needs to hedge assets and liabilities as part of their basic business, and is subject to movements in the price moves of various underlying assets as well as currencies and rates.
4. **Importer:** Consider a goods importer who is subject to the prices of goods priced in a different currency.
5. **Operational:** Suppose we find we had less inventory of an asset than previously accounted for. We are now subject to the vagaries of the market to replace the inventory.

6. **Accounting:** Derivatives are accounted on a mark-to-market basis using traded market values or models that reflect the so-called fair value. Any changes in basic inputs could, therefore, lead to profit or losses.

7. **Real Options:** We discussed so-called real options in Section 4.4. If we made an investment in a venture that was valued as a real option, we stand to make or lose money if some basic economic variables changed.

It should be clear, therefore, that if we understand the market risks inherent in our company that we are placing ourselves in a position to hedge or mitigate these risks, or even gear up on these risks!

18.3 RISK MANAGEMENT PROCESS

Our total market risk management ethos consists of the following key elements:

1. **Risk Identification:** It should come as no surprise that the first element of our management programme is the identification of all risks.

    This means a thorough understanding of our business, and proper examination of our income statement and balance sheet, is needed to understand which market variables give rise to changes in these. Ideally, this is done by examination of periodic management accounts, such as daily P/L statements for the overall business, and comparing those with changes in underlying market variables.

2. **Risk Measurement:** Having identified sources of risk, it becomes crucial to quantify the extent of such risk.

    This is usually done through several risk measurement metrics that range from simple spot equivalent positions to VaR and stress testing, as examples, depending on the sophistication of the company involved in the analysis. Let us consider these in a bit more detail:

    a. **Risk Proxies:** If a company is exposed to a set of risks, which are very similar, a proxy could be made to perform a 'back-of-the-envelope' calculation to quantify the risk. An example might be to approximate a basket of equities by an index. The advantage of such an approach is simplicity and yielding an 'effective' position that management relates to. The obvious disadvantage is that we are making some serious assumptions with regard to basis risks between different (but related) positions.

These measures are also frequently applied at a global level where we would look at the overall exposure, say the amount of exposure to interest rates increasing by a basis point. This would look across the whole yield curve and, again, ignore the fact that different areas of the curve do not always move by the same amount.

b. **Value-at-Risk (VaR):** Value-at-Risk (VaR) presents another measure of market risk frequently used by banks and regulators. VaR has been defined as the loss (stated with a specified probability) from adverse market movements over a fixed-time horizon, assuming the portfolio is not managed during this time.

So, given a fixed time horizon, and a predetermined confidence level, say α, the value-at-risk, VaR_α is the loss in the market value of the portfolio, over the time horizon, that is exceeded with probability $1 - \alpha$, that is,

$$Pr(\delta P < \text{VaR}_\alpha) = 1 - \alpha,$$

where we denote δP as the change in the portfolio under consideration over the time horizon. See, [42].

Hence, VaR is measured as the lower percentile of a distribution for theoretical P/L that arises from possible movements of the market risk factors over a fixed-time horizon. See, [82, 106], for example.

c. **Stress Tests:** Stress tests typically consider the effect of some hypothetical scenario applied to our overall portfolio. An example could be to see how much we make/lose if the exchange rate depreciates by 20% overnight. Stress tests are very useful to identify risk but frequently also ignore basis risks. The results from stress tests could also be ignored because management finds the scenario totally implausible.

The advice should be that the quantum of the P/L effect is not as important as the identification of exposure that we have. Stress tests could be very useful to identify embedded optionality.

It is important to understand that no risk measurement metric is perfect – the interpretation thereof is very important, though, as well as how we put it to further use.

3. **Monitoring and Control:** If we are able to attach a metric to an identified risk, it means that we have a sense of the economic damage that could be caused as a result of that risk. We should judge the economic benefit from having the risk – such as the amount of profit or other advantage that we would derive – against that.

   Typically, we would not want to have unbounded amounts of any risk(s); consequently, we would impose limits to control these risks. It is, therefore, important to ensure that we set our controls (limits) to be reflective of the identified risks, and that these accurately control the exposure that will ultimately feed through to our income statement.

   Therefore, our process is to monitor the measured risks against limits on a periodic basis and to ensure the correct level of control of our risks. This needs to be tested against the set levels of our risk-appetite.

4. **Testing:** Market risk management is most successful in an environment where the risks are transparently discussed and understood. It is of the utmost importance to dissect the risks and to understand whether our understanding of its impact to the balance sheet and income statement is correct, whether our measurement is correct, and whether we are adequately compensated for the risk.

   Part of the process of challenging and testing risk measurements would be to ensure that we have captured all our risks. We continually need to revisit all our assumptions throughout the whole process. We started off by investigating our income statement – it is, therefore, only fair that we return there. Our risk measurements must balance back to the income statement in the sense that market moves applied to market sensitivities should lead to P/L numbers; if this is not the case, it probably means that we are missing something from our risk model or it raises the possibly of some operational error, for example.

   We need to ensure that we understand and question hidden risks. These are frequently basis risks; for example, being long a position in a government bond against a paid position in a swap. On an overall interest rate risk basis, we could have no exposure as long as both instruments move in tandem, but the two underlying base curves could move in different ways to create the basis risk. However, elementary measurement models would not always capture these type of risks.

18.4 POTENTIAL PROBLEMS

There are many pitfalls in market risk management. As an example, when identifying risks, we often have incomplete P/L information resulting from inadequate systems and processes. When measuring risks, we frequently encounter insufficient data or inadequate proxies – typically when valuing derivative-based contracts.

While systems and processes can always be improved, it is of the utmost importance that we have the ability to interpret results. Management can sometimes be impressed by very sophisticated concepts; yet, in a crunch-time when heavy-hitting decisions need to be taken, the simple concepts often provide the most intuition and management information.

18.5 CONCLUSION

This chapter is effectively a coffee-break discussion on market risks and the management thereof. There are significant benefits from managing market risk within a proper framework. Particularly, a real understanding of the cause and effect of financial market forces on our financial results.

It is fitting to end this chapter with the insightful remarks of Don Chance [25]:

> Questions are always asked when money is lost. Most organizations and certainly most treasurers and traders never see the inconsistency in making huge sums of money in a highly efficient and competitive market where gains equal losses. Success is equated with making money and to question success is to raise suspicions that talent was not involved. To avoid this problem a firm must adjust profits for the risk taken. Firms that do this are asking the right questions.

References

[1] M. Abramowitz and I.A. Stegun, *Handbook of Mathematical Functions.* Dover, 1965.

[2] Y. Achdou and O. Pironneau, *Computational Methods for Option Pricing.* Society for Industrial and Applied Mathematics, SIAM, 2005.

[3] C. Albanese and G. Campolieti, *Advanced Derivatives Pricing and Risk Management: Theory, Tools, and Hands-on Programming Applications.* Elsevier, 2005.

[4] L. Bachelier, M. Davis, A. Etheridge and P.A. Samuelson, *Louis Bachelier's Theory of Speculation.* Translated by Mark Davis & Alison Etheridge, Princeton University Press, 2006.

[5] Bank for International Settlement, *OTC Derivatives Statistics at End-June 2023*, BIS, 16 November 2023.

[6] M.W. Baxter and A.J.O. Rennie, *Financial Calculus: An Introduction to Derivative Pricing.* Cambridge University Press, 2002.

[7] P.L. Bernstein, *Against the Gods: The Remarkable Story of Risk.* Wiley, 1996.

[8] F. Black, The Pricing of Commodity Contracts, *Journal of Financial Economics*, 3(1-2), pp. 167-179, 1976.

[9] F. Black and M. Scholes, The Pricing of Options and Corporate Liabilities, *Journal of Political Economy*, 81, pp. 637-654, 1973.

[10] A.S. Blinder, *Central Banking in Theory and Practice.* MIT Press, 1999.

[11] S. Blyth, *An Introduction to Quantitative Finance.* Oxford University Press, 2014.

[12] D.T. Breeden, and R.H. Litzenberger, Prices of State-contingent Claims Implicit in Option Prices, *Journal of Business*, Vol. 51, pp. 621-651, 1978.

[13] Board of Governors of the Federal Reserve System, SR 11-7 *Guidance on Model Risk Management.* Board of Governors of the Federal Reserve System, 2011.

[14] J.-P. Bouchaud and M. Potters, *Theory of Financial Risk and Derivative Pricing*, 2nd ed. Cambridge University Press, 2003.

[15] P.P. Boyle, Options: A Monte Carlo Approach, *Journal of Financial Economics*, 4(3), pp. 323-338, 1977.

[16] M. Broadie and J.B. Detemple, Option Pricing: Valuation Models and Applications, *Management Science*, 50(9), pp. 1145-1177, 2004.

[17] M. Broadie and P. Glasserman, Estimating Security Price Derivatives Using Simulation, *Management Science*, 42(2), pp. 269-285, 1996.

[18] M. Broadie, P. Glasserman and S. Kou, A Continuity Correction for Discrete Barrier Options, *Mathematical Finance*, 7(197), pp. 325–348.
[19] T. Björk, *Arbitrage Theory in Continuous Time*. Oxford University Press, 1998.
[20] A.J.G. Cairns, *Interest Rate Models: An Introduction*. Princeton University Press, 2004.
[21] P. Carr, H. Geman, D.B. Madan and M. Yor, Stochastic Volatility for Lévy Processes, *Mathematical Finance*, 13, pp. 345–382, 2003.
[22] P. Carr and D. Madan, Towards a Theory of Volatility Trading. In R. Jarrow (ed.), *Risk Book on Volatility*, Risk, 1998, pp. 417–427.
[23] P. Carr and L. Wu, Static Hedging of Standard Options, *Journal of Financial Econometrics*, 12(1), pp. 3–46, 2014.
[24] J. Cassidy, *How Markets Fail: The Logic of Economic Calamities*. Farrar, Straus and Giroux, 2009.
[25] D.M. Chance, *Essays in Derivatives*. Frank Fabozzi Associates, 1998.
[26] D.M. Chance, A Synthesis of Binomial Option Pricing Models for Lognormally Distributed Assets, *Journal of Applied Finance*, 18(1), p. 38, 2008.
[27] E. Chancellor, *The Price of Time: The Real Story of Interest*. Penguin, 2022.
[28] N. Chriss, *Black-Scholes and Beyond: Option Pricing Models*. McGraw-Hill, 1997.
[29] L. Clewlow and C. Strickland, *Implementing Derivatives Models*. Wiley Financial Engineering, 1998.
[30] R. Cont and P. Tankov, *Financial Modelling with Jump Processes*. Chapman & Hall/CRC, 2004.
[31] U.S. Congress, *JP Morgan Chase Whale Trades: A Case History of Derivatives Risks and Abuses*. Congress: Senate, 2013.
[32] R. Cont and P. Tankov, Constant Proportion Portfolio Insurance in the Presence of Jumps in Asset Prices, *Mathematical Finance: An International Journal of Mathematics, Statistics and Financial Economics*, 19(3), pp. 379–401, 2009.
[33] J. Cox and S. Ross, The Valuation of Options for Alternative Stochastic Processes, *Journal of Financial Economics*, 3, pp. 145–166, 1976.
[34] J.C. Cox, S.A. Ross and M. Rubinstein, Option Pricing, a Simplified Approach, Vol. 7, *Journal of Financial Economics*, pp. 229–263, 1979.
[35] M.H. Davis, *Black–Scholes Formula, Encyclopedia of Quantitative Finance*, pp. 199–207, New York: John Wiley, 2010.
[36] P. Diaconis and B. Skyrms, *Ten Great Ideas About Chance*. Princeton University Press, 2018.
[37] E. Derman, *My Life as a Quant: Reflections on Physics and Finance*. Wiley, 2004.
[38] E. Derman, *Models Behaving Badly: Why Confusing Illusion with Reality can Lead to Disaster, on Wall Street and in Life*. Simon and Schuster, 2011.
[39] E. Derman and I. Kani, Riding on a Smile, *Risk*, 7(2), pp. 32–39, 1994.
[40] E. Derman and M.B. Miller, *The Volatility Smile*, John Wiley & Sons, 2016.
[41] F. de Weert, *Exotic Options Trading*, Wiley Finance, 2008.

[42] P. Embrechts, R. Frey and A. McNeil, *Quantitative Risk Management*, Princeton University Press, 2011.
[43] B. Dupire, Pricing with a Smile, *Risk*, 7(1), pp. 18–20, 1994.
[44] Franklin R. Edwards and Michael S. Canter, The Collapse of Metallgesellschaft: Unhedgeable Risks, Poor Hedging Strategy, or Just Bad Luck?, *Journal of Applied Corporate Finance*, 8, Spring, p. 211–264, 1995.
[45] A. Etheridge, *A Course in Financial Calculus*. Cambridge University Press, 2002.
[46] M. Evans, N. Hastings and B. Peacock, *Statistical Distributions*, 2nd ed. Wiley, 1993.
[47] Review of the Federal Reserve Supervision and Regulation of Silicon Valley Bank. April 2023, https://www.federalreserve.gov/publications/2023-April-SVB-Key-Takeaways.htm.
[48] J. Fouque, G. Papanicolaou and K. R. Sircar, *Derivatives in Financial Markets with Stochastic Volatility*. Cambridge University Press, 2001.
[49] L.V. Gave, *Too Different for Comfort*. GaveKal Books, 2013.
[50] J. Gatheral, *The Volatility Surface: A Practitioners Guide*, Vol. 357. John Wiley and Sons, 2011.
[51] R. Geske, The Valuation of Corporate Liabilities as Compound Options, *Journal of Financial and Quantitative Analysis*, 12, pp. 541–552, 1979.
[52] F. Guillaume and W. Schoutens, Calibration Risk: Illustrating the Impact of Calibration Risk under the Heston Model, *Review of Derivatives Research*, 15, pp. 57–79, 2012.
[53] A. Green, *XVA: Credit, Funding and Capital Valuation Adjustments*. John Wiley & Sons, 2015.
[54] D.J. Higham, Nine Ways to Implement the Binomial Method for Option Valuation in Matlab, vol 44(4), *SIAM Review*, pp. 661–677, 2002.
[55] D.J. Higham, *An Introduction to Financial Option Valuation: Mathematics, Stochastics and Computation*, Vol. 13. Cambridge University Press, 2004.
[56] T. Hoggard, A. E. Whalley and P. Wilmott, Hedging Option Portfolios in the Presence of Transaction Costs, *Advances in Futures and Options Research*, Emerald Publishing, 1993.
[57] L.P. Hughston and C.J. Hunter, *Financial Mathematics: An Introduction to Derivatives Pricing*, Lecture Notes. King's College London, 2000.
[58] J.C. Hull, *Options, Futures and Other Derivatives*, 5th ed. Prentice Hall, 2003.
[59] C.B. Hunzinger and C.C. Labuschagne, Pricing a Collateralized Derivative Trade with a Funding Value Adjustment, *Journal of Risk and Financial Management*, 8(1), pp. 17–42, 2015.
[60] P. Jäckel, *Monte Carlo Methos in Finance*. Wiley Finance Series, 2002.
[61] R.A. Jarrow and S.M. Turnbull, *Derivative Securities*. International Thomson Publishing, 1996.
[62] P. Jorion, Risk Management Lessons from Long-Term Capital Management, *European Financial Management*, 6, pp. 277–300, 2000.
[63] M.S. Joshi, *The Concepts and Practice of Mathematical Finance*. Cambridge University Press, 2003.

[64] M. King, *The End of Alchemy: Money. Banking, and the Future of the Global Economy*. Little, Brown Book Group, 2016.

[65] Y.K. Kwok, *Mathematical Models of Financial Derivatives*. Springer, 2008.

[66] D.P.J. Leisen and M. Reimer, Binomial Models for Option Valuation - Examining and Improving Convergence, *Applied Mathematical Finance*, 3, pp. 319–346, 1996.

[67] H.E. Leland, Option Pricing and Replication with Transaction Costs, *Journal of Finance*, 40, pp. 1283–1301, 1985.

[68] M. Lewis, *Liar's Poker*. Hodder & Staunton, 1989.

[69] A.W. Lo, *Adaptive Markets: Financial Evolution at the Speed of Thought*. Princeton University Press, 2017.

[70] J. Loeys, What Have I Learned? *Global Asset Allocation, The J.P. Morgan View*, JP Morgan, November 2017.

[71] R. Lowenstein, *Origins of the CRASH*. The Penguin Press, 2004.

[72] D.G. Luenberger, *Investment Science* Oxford University Press, 1998.

[73] E. Lefevre, *Reminiscences of a Stock Operator*. John Wiley & Sons, 2010.

[74] D.B. Madan, On the Modelling of Option Prices, *Quantitative Finance*, 1(5), p. 481, 2001.

[75] B.G. Malkiel, *A Random Walk Down Wall Street: The Time-Tested Strategy for Successful Investing*. WW Norton & Company, 2012.

[76] H. Marks, *The Indispensability of Risk*. Oaktree, April 2024.

[77] L. Martellini, P. Priaulet and S. Priaulet, *Fixed-income Securities: Valuation, Risk Management and Portfolio Strategies*, Vol. 237. John Wiley & Sons, 2003.

[78] M.J. Mauboussin, *More Than You Know*. Columbia Business School, 2008.

[79] R.C. Merton, Rational Theory of Option Pricing, *Bell Journal of Economic and Management Science*, 4, pp. 141–183, 1973.

[80] R.C. Merton, On the Pricing of Corporate Debt: The Risk Structure of Interest Rates, *Journal of Finance*, 29, pp. 449–470, 1974.

[81] R.C. Merton, Option Pricing When Underlying Stock Returns are Discontinuous, *Journal of Financial Economics*, 3, pp. 125–144, January-February 1976. (Chapter 9 in Continuous-Time Finance.)

[82] Jorge Mina and Jerry Yi Xiao, *Return to RiskMetrics: The Evolution of a Standard*. RiskMetrics Group, April 2001.

[83] M. Musiela and M. Rutkowski, *Martingale Methods in Financial Modelling*. Springer, 1998.

[84] S.N. Neftci, *An Introduction to the Mathematics of Financial Derivatives*. Academic Press, 1996.

[85] D.B. Nelson and K. Ramaswamy, Simple Binomial Processes as Diffusion Approximations in Financial Models, *The Review of Financial Studies*, 3, pp. 393–430, 1990.

[86] D. O'Kane, *Modelling Single-name and Multi-name Credit Derivatives* John Wiley and Sons, 2008.

[87] S.C. Pirrong, Metallgesellschaf: A Prudent Hedger Ruined, or a Wildcatter on Nymex?, *The Journal of Futures Markets*, 17(5), pp. 543–578, 1997.

[88] V. Piterbarg, Funding beyond Discounting: Impact of Stochastic Funding and Collateral Agreements and Derivatives Pricing, vol 23, pp 97–102, *Risk*, 2010.

[89] E. Platen and M. Schweizer, On Feedback Effects from Hedging Derivatives, *Mathematical Finance*, 8, pp. 67–84, 1998.

[90] A.D. Polyanin, *Handbook of Linear Partial Differential Equations for Engineers and Scientists*. Chapman and Hall/CRC, 2001.

[91] W.H. Press, S.A. Teukolsky, W.T. Vetterling and B.P. Flannery, *Numerical Recipes in C: The Art of Scientific Computing*, 2nd ed. Cambridge University Press, 1992.

[92] R. Rebonato, *Volatility and Correlation: The Perfect Hedger and the Fox*. John Wiley & Sons, 2005.

[93] P. Ritchken, *Derivative Markets: Theory, Strategy and Applications*. HarperCollins, 1996.

[94] N. Sauer, Energy Methods and the Black-Scholes Partial Differential Equation, Technical Report UPWT 2009/1, January 2009.

[95] P.J. Schönbucher and P. Wilmott, The Feedback Effect of Hedging in Illiquid Markets, *SIAM Journal on Applied Mathematics*, 61(1), pp. 232–272, 2000.

[96] W. Schoutens, *Lévy Processes in Finance: Pricing Financial Derivatives*. Wiley, 2003.

[97] W. Schoutens, E. Simons and J. Tistaert, A Perfect Calibration! Now What? *The Best of Wilmott*, p. 281, Wilmott Magazine, 2003.

[98] M. Schroder, Changes of Numeraire for Pricing Futures, Forwards, and Options, *The Review of Financial Studies*, 12(5), pp. 1143–1163, 1999.

[99] F. Schwed, *Where are the Customers' Yachts?, or, A Good Hard Look at Wall Street*. Wiley Investment Classics, 1940.

[100] N. Taleb, *Dynamic Heding: Managing Vanilla and Exotic Options*. Wiley Financial Engineering, 1997.

[101] G. Tassey, The Economic Impacts of Inadequate Infrastructure for Software Testing. National Institute of Standards and Technology, Report 6, 2002.

[102] U.S. Financial Crisis Inquiry Commission, Final Report of the National Commission on the Causes of the Financial and Economic Crisis in the United States, 2011.

[103] P. Wilmott, *Frequently Asked Questions in Quantitative Finance*. John Wiley & Sons, 2010.

[104] P. Wilmott, *Paul Wilmott on Quantitative Finance*. John Wiley & Sons, 2013.

[105] P. Wilmott, S. Howison and J. Dewynne, *The Mathematics of Financial Derivatives: A Student Introduction*. Cambridge University Press, 1995.

[106] Yasuhiro Yamai and Toshinao Yoshiba, *On the Validity of Value-at-Risk: Comparative Analyses with Expected Shortfall, Monetary and Economic Studies*. Monetary and Economic Studies, Bank of Japan, January 2002.

Index

A

account, 186
accuracy, 174, 176
American, 43, 91, 135
analytic, 176
approximation, 61, 66, 83, 101, 167, 168, 174, 176
arbitrage, 7, 23, 28, 39, 62, 65, 66, 77, 78, 89, 115, 168
asset, 13, 21, 43, 51, 140
assumption, 61
assumptions, 82
at-the-money, 106
average, 156

B

balance sheet, 61, 65, 185
barrier, 134, 159, 165
basis risk, 189
basket, 161
binary, 154, 159
binomial, 77, 78, 81, 85, 86, 88, 90, 92, 99, 111, 112
Black, 105, 126
Black Box, 178
Black–Scholes, 128
Black–Scholes, 50, 86, 100, 103, 112, 113, 132, 138, 140, 167
BOJ, 32
bond, 7, 13, 32, 38
bootstrap, 40
bound, 61, 64, 71
boundary conditions, 124, 133
butterfly, 56, 72

C

Calendar, 56
calibration, 164, 173
call, 62, 66, 71, 79, 88, 100, 101, 103, 133, 167
capital structure, 140
carry, 44
Cauchy-Euler, 132
central bank, 12, 29–31, 40
Charm, 146
claim, 44, 88, 101
cliquet, 158
closed-form, 144, 167, 176, 177
CME, 6
collar, 57
collateral, 119
Colour, 146
commodities, 1, 7, 10, 185
compensate, 188
compound option, 159
compounding, 38
consistent, 86
contingent, 99, 101
contingent claim, 44, 61, 74, 81
continuous, 64
contract, 42, 126
convex, 71
cost, 44, 49, 55, 65
CPPI, 118, 119

credit, 114, 140
crises, 30
currency, 1, 11, 31

D

debt, 13, 138–141
default, 38, 81, 137, 140
Delta, 145, 149, 150
derivative, 6–8, 74, 101, 123, 124, 148, 149, 173
Derman, 60
differential equation, 132
differentiate, 73, 150
digital, 154
discounted, 68
discretization, 85
distribution, 57, 74, 116, 167, 174
dividend, 61, 116
dual, 128
dynamic hedging, 175
dynamics, 99, 113

E

ECB, 32
embedded, 45, 187
Enron, 182
equity, 1, 8, 185
error, 170, 176
ETF, 7, 9, 33
Euler, 65
European, 43, 50, 62, 66, 68, 71, 91, 99, 103, 128, 133, 134
exchange, 4, 7, 8, 43
exercise, 43, 46, 68, 71, 128, 135, 172
exotic, 150, 163, 174
exotic option, 156
expected, 99, 101
expected value, 77
expiry, 43, 174
exposure, 68

F

fair value, 173
FED, 32
financial, 12
financial engineering, 184
fluctuation, 2, 21

forex, 7, 11, 185
forward, 105
forward contract, 42, 104, 125
forward starting option, 103, 157
frictions, 61
function, 73
functional, 40
fund, 23
funding, 14
futures, 5, 6, 8, 14, 43, 126

G

Gamma, 146
gap risk, 119
gas, 9
GFC, 182
goal, 22
gold, 9
government, 40
Greeks, 67, 144, 147, 150

H

hedge, 22, 79, 81, 150
hedge funds, 7, 27
hedging, 22, 26, 115, 120
homogeneous, 65
horizon, 21

I

identification, 186, 188
illiquid, 25
impact, 113
implement, 169
implied volatility, 53
income statement, 185
index, 50
indices, 8
inflation, 31
initial condition, 124
input, 171, 173
instruments, 3, 14
insurance, 68
integral, 73, 102
integration, 74
interest, 13, 37, 39
interpolation, 40
intrinsic, 43, 92

investor, 10
issuer, 13

J

jumps, 113, 118

L

liability, 68
limits, 188
linear, 73
Lipschitz, 64
liquidate, 140
liquidation, 115
liquidity, 12, 26, 61
loan, 66
local volatility, 128
lookback, 156, 157
loss, 181
LTCM, 181

M

Maclaurin, 100, 101, 111
macro, 28
margin, 43, 115
market, 2, 6, 19, 22, 23, 26, 77, 181
market risk, 184, 186
maturity, 38, 43, 63, 64, 72
mean, 99
measurement, 186
Merton, 50, 140
Metallgesellschaft, 181
model, 50, 60, 81, 169, 171, 172, 174, 179
monetary policy, 33
money market, 39, 81
monitoring, 188
Monte Carlo, 47, 106–108
multi-asset, 161
multiplier, 119

N

Nelson-Siegel, 40
neutral, 114
NIRP, 33
notes, 14
notional, 38
numerical, 174

O

obligation, 43, 44
oil, 9
option, 8, 14, 43, 46, 50, 55, 61, 62, 66, 67, 73, 100, 102, 132, 133, 137, 167
optionality, 45, 187
order, 100
OTC, 42
output, 172

P

paradigm, 173, 174
parameters, 167
parametric, 40
parity, 68
partial differential equation, 86, 124
payoff, 43, 55, 57, 74, 92, 99, 123
PDE, 86, 113, 116, 124, 128, 132, 147
pension, 7
perceptions, 2
perpetual, 132–134
physical, 8
policy, 13
polynomial, 167
portfolio, 66, 69, 72, 112, 114, 187
premium, 44
price, 2, 8, 10, 65, 101, 113
pricing, 74
probability, 77, 79
process, 101, 113, 172, 179
producer, 23
provision, 171
put, 44, 66, 67, 71, 91, 103
put-call parity, 66, 69, 91, 103, 139

Q

quadrature, 108
quantitative, 28
quantitative easing, 32
quanto, 155, 162

R

rainbow option, 161
rate, 33, 38–40, 101
rates, 13, 30, 67, 185
real, 47

real option, 45
realized volatility, 52
rebalance, 114
rebate, 159
recombining, 83, 92
recovery, 140
recursion, 88, 90, 92
regulation, 34, 36
regulator, 34, 36
replicate, 79, 121, 176
repurchase, 120
Rho, 145
right, 43, 44, 49
risk, 14, 23, 169, 179, 181, 186–189
risk management, 184
risk-free, 13, 112, 155
risk-neutral, 74, 79, 81, 99, 101, 123, 130

S

security, 2
sensitivity, 148
sentiment, 12
series, 100
shares, 7
simulation, 176
solution, 133, 176
sovereign, 13
speculator, 7, 20
speed, 146
spot, 62, 155
spread, 56, 72, 141
stability, 22
standard error, 108
state, 61
stock, 7, 61, 78, 81, 99, 101
stock exchange, 4
straddle, 55
strangle, 56
strategies, 55
stress test, 187
strike, 64, 65, 68, 71, 155
structure, 12
subjective, 77
surface, 150, 164

survival, 140
symmetry, 126
synthetic, 66

T

taxes, 61, 65
Taylor, 85, 90, 105, 112, 124, 144
Taylor's theorem, 83, 86
tenor, 155
term-structure, 55
Theta, 132, 145
trade, 11, 12, 26
trading, 7
transaction, 65
transformation, 129
tree, 88, 91, 92
trend, 99
turnover, 19

U

underlying, 61, 79
unwind, 61

V

validation, 172, 176
value, 2, 73, 79
Vanna, 146
VaR, 187
variability, 51
variance, 55, 82, 83, 99
Vega, 149
verification, 176
volatility, 10, 22, 50–52, 55, 75, 115, 128, 141, 155, 159, 185
Volga, 146
volume, 12
vomma, 146

Y

yield curve, 40

Z

ZCB, 38, 139
zero cost, 57
zomma, 147

Milton Keynes UK
Ingram Content Group UK Ltd.
UKHW020759121224
451979UK00021B/37